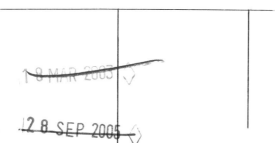

Industrial Biocides
Selection and Application

Industrial Biocides
Selection and Application

Edited by

D.R. Karsa
Akcros Chemicals UK Ltd, Manchester, UK

D. Ashworth
BASF Biocides Limited, Nottingham, UK

ROYAL SOCIETY OF CHEMISTRY

The Proceedings of the Biocides Agenda '99 meeting held on 14–16 September 1999 at the University of Salford, UK.

Special Publication No. 270

ISBN 0-85404-805-7

A catalogue record for this book is available from the British Library

Published by The Royal Society of Chemistry,
Thomas Graham House, Science Park, Milton Road,
Cambridge CB4 0WF, UK
Registered Charity No. 207890

For further information see our web site at www.rsc.org

Printed by Athenaeum Press Ltd, Gateshead, Tyne and Wear, UK

Preface

This book contains the proceedings of an international conference held in Salford, England, in the Autumn of 1999. The aim of the conference, reflected in the papers within the book, was to provide practical guidance on the use of biocides across a broad range of applications. The backdrop for the conference was set by the publication in Europe of the Biocidal Products Directive, which provided both a regulatory framework and the segregation of the market into defined application sectors. This book describes the overall biocides market and the consequences of the implementation of the Biocidal Products Directive. It gives practical advice on the specific use of biocides in key sectors. Certain key application areas such as Paper, Water, Coatings and Disinfection are covered in considerable depth, reflecting diverse employment of biocides in these industries. Other more tightly defined applications, such as the use of biocides in swimming pools and portable toilet fluids, are looked at in detail. The book has the level of detail required by skilled technical personnel whilst at the same time providing a broad overview relevant to more commercially oriented practitioners looking to develop their knowledge base.

Contents

Disinfection

Plastics

Performance Fluids

Swimming Pools

Pharmaceuticals, Cosmetics and Toiletries

Overview

AN OVERVIEW OF THE BIOCIDES MARKET

Niall D'Arcy

The Biocides Information Service, University College, Dublin, Eire

SUMMARY

By definition, biocides are used to kill or inhibit the growth of living organisms in industrial and consumer products.
   There are five classes of biocide:

Bacteriostats:	To control bacteria on skin and surfaces
Bactericides:	To protect from bacterial spoilage
Fungicides:	To protect from fungal attack and spoilage
Insecticides:	To protect against insects
Algaecides:	To protect against algae

The value ($M) of the speciality global biocides market by application sector is given in Table I.

Table I: *Speciality Biocides ($M) at A.I. Level*

Industry	N America	Europe	Japan	Asia-Pacific	Latin America	Total
Personal Care	130	110	50	20	20	**330**
Household products	50	60	25	10	10	**155**
Coatings	150	100	60	40	20	**370**
Metalworking fluids	50	40	30	20	10	**150**
Paper	50	40	30	20	20	**160**
Wood	300	160	60	80	60	**660**
Plastics	50	20	20	20	10	**120**
Disinfection	100	60	40	40	20	**300**
Textiles	20	10	10	10	5	**55**
Water Treatment	280	130	70	30	20	**530**
Total	**1,180**	**730**	**495**	**290**	**195**	**2,790**

The current biocide industry trends are dominated by increased regulatory and environmental pressures from:
- Biocidal Products Directive/EPA/MITI.
- Legislation for "new" sectors i.e. cutting boards
- Pressure on chlorinated actives/tin based products
- Eco-labeling, e.g. VOC free products
- Alternatives to formaldehyde release chemicals

Market and competitive trends include moves towards:
- Water-based products
- Preservation, to cut disposal costs
- Global supply agreements
- Patent expiry; increased competitiveness
- Industry consolidation
- Consumer awareness of "germs"
- High future growth in emerging markets, e.g. Asia

The Biocides Market and End-user Requirements
The split in the biocides market is given in Table II.

Table II: *Distribution in the Biocides Industry*

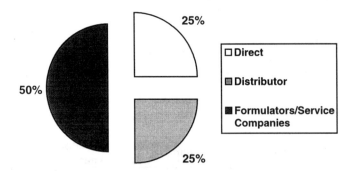

The end-users are seeking biocides which are:

- Effective against relevant micro-organisms
- Cost effectiveness
- Broad regulatory approval (BGA, MITI, FDA)
- Externally tried and tested product
- Strong toxicology package
- Formulation compatibility
- Wide pH stability
- Good temperature stability

- Solvent free formulation
- High quality product, minimal impurities
- No skin sensitization
- Biodegradable

while the end-users are seeking the following from their suppliers:

- Products of consistent quality (ISO 9002)
- Guarantee security of supply
- Access to competent technical personnel that offer regulatory, technical, formulation and safety advice
- Broad regulatory approval for products
- Reliable, safe handling services.

The 'Biocides Information Services' (www. biocide-information.com) and 'Biocides Reporting Service' are based in Dublin/Eire at University College.

Biocides Reporting Service provides <u>monthly</u> information on biocides and end-user industries through the world. The information searches are tailored to individual business needs and sent to key members of that company. The information is sourced from over 200 worldwide trade journals and covers: competitors, customers, industry analysis & trends, regulations and patents.

The skills of the Biocide Information Service include:

- Multidisciplined professionals
- Extensive databases
- Language fluency
     English, French, German, Italian, Spanish

with market expertise in applications such as:

     Coatings, Disinfectants, Household products, Ink, Metalworking Fluids, Personal care, Plastics, Water treatment, Wood

and business expertise by way of:

     Customer satisfaction, Concept testing, Distribution, Pricing

The Biocides Reporting Service functions by providing:

- Global Biocides Directory

- Multiclient Studies
     Biocides in Western Europe
     Biocides in North America
     Biocides in Asia Pacific (Future)

- Single Client Research
     Market Analysis

The Political Arena

THE BIOCIDAL PRODUCTS DIRECTIVE (98/8/EC)

Geoffrey Wilson

Health & Safety Executive, Rose Court, 2 Southwark Bridge, London SE1 9HS

1 SUMMARY

This paper outlines the content of the Biocidal Products Directive and explains how it will be implemented in the UK. It covers what industry will have to do in compliance, what is currently happening in Europe and gives an estimated timetable for all the activities.

2 BIOCIDAL PRODUCTS DIRECTIVE

A proposal for a European Parliament and Council Directive concerning the placing of biocidal products on the market was adopted on 16 February 1998, numbered 98/8/EC and published in the Official Journal on 24 April 1998 (reference L 123 pp 1-62).
  The Directive has to be implemented in all member states by 14 May 2000.

2.1 Aims of the Directive

The Directive is based on article 100a of the Treaty and it has has two key aims:
- HARMONISATION OF THE MARKET FOR BIOCIDAL PRODUCTS - A HIGH LEVEL OF PROTECTION FOR HUMANS AND THE ENVIRONMENT

2.2 Operation of the Directive

The Directive will operate by listing all active substances which can be used in biocidal products in a list (annex I to the Directive) and requiring that only those active substances listed can be used in biocidal products. Member states will then authorise biocidal products to a set of common principles (annex V of the Directive) with a system of mutual recognition of authorisations.

2.3 Content of the Directive

A summary of the main contents of the Directive is:

MUTUAL RECOGNITION
SCOPE - there are 23 Product types
MUTUAL RECOGNITION
COMPARATIVE ASSESSMENT
DATA REQUIREMENTS
DATA PROTECTION & CONFIDENTIALITY
SIMPLIFIED PROCEDURES
CLASSIFICATION, PACKAGING and LABELLING
CHARGES
TRANSITIONAL ARRANGEMENTS
GUIDANCE DOCUMENTS

Member states have a number of actions under the Directive. They have to appoint a competent authority (or authorities), transpose the Directive into national law, enforce it and provide information to the European Commission.

3 IMPLEMENTATION OF THE BPD IN THE UK

The UK has set three principles for the implementation of the Directive. These are that the implementing systems must be:
- EFFICIENT & EFFECTIVE
- TRANSPARENT & ACCOUNTABLE
- SELF FINANCING (as far as possible).
In practice it has been agreed that the Health and Safety Executive (HSE) will operate as the de facto competent authority and that new Regulations will be made under the Health & Safety at Work etc. Act and the European Communies Act - the Biocidal Products Regulations (BPR).

Guidance on the new Regulations and on the proposed new systems will also be produced and both the Regulations and the guidance will be circulated for public comment. A number of industry groups have been set up to assist with this process. One group deals with financial issues, another with the transitional arrangements whilst a third group covers optimisation of use - this latter group is known as the Biocides Users Group (BUG).

There is a current control system for certain types of biocidal products (known as non-agricultural pesticides) operated by HSE under the Control of Pesticides Regulations (CoPR). These are antifouling products, insecticides, wood preservatives and surface biocides. Additionally agricultural pesticides are controlled under CoPR but these are administered by the Pesticides Safety Directorate (PSD), an executive agency of the Ministry of Agriculture, Fisheries and Food (MAFF).

Many new product types will be regulated by HSE under the BPR, and these are listed on the next page:

DISINFECTANTS
PRESERVATIVES - In can; Film; Fibre, rubber & polymerised; Liquid cooling   & processing; Metalworking; Food & feedstocks preservatives.
SLIMICIDES
EMBALMING & TAXIDERMIST FLUIDS
ATTRACTANTS
(Rodenticides, Avicides, Molluscicides, Piscicides, Vertebrate control agents).

The last group of products - enclosed in brackets - are currently controlled under CoPR but by PSD rather than HSE.

3.1 Proposals for charges

The Directive (article 25) allows for full cost recovery and HSE's proposals are that this will be collected in two ways: by fees and by a general industry charge (GIC). A fee will be charged to directly identifiable customers e.g. those who apply for authorisation of a biocidal product whilst the GIC will be payable by all with biocidal products on the market. It will cover the costs of work such as monitoring, specific biocides research work etc.; it is likely to start on 14 May 2000.

3.2 The Biocides Consultative Committee (BCC)

This committee will offer advice to the competent authority (HSE) on all aspects involving biocides. It will be composed mainly of independent members and have an independent chairman. Other Government Departments will sit on the committee acting as advisors. There will also be an interdepartmental committee which will work with the competent authority on the scientific assessments and risk assessments.

3.3 Guidance Documents

Five sets of guidance will be produced:
GUIDANCE FOR SUPPLIERS
GUIDANCE FOR USERS
TRANSITIONAL GUIDANCE
GUIDANCE FOR APPLICANTS
A SIMPLE GUIDE
This guidance should all be available before the implementation date of the Directive, with the Transitional Guidance being available first.

3.4 Public Consultation

A Consultative Document (CD) containing:
- THE DRAFT REGULATIONS
- THE BIOCIDAL PRODUCTS DIRECTIVE
- THE DRAFT GUIDANCE
- THE REGULATORY IMPACT ASSESSMENT
- A LIST OF CONSULTEES
- A REPLY PRO-FORMA
was published on 2 July 1999 and comments were invited from industry, other government departments, non-governmental organisations and the general public up to 1 October 1999.

After this date the draft Regulations and Guidance will be amended in the light of comments received and then a submission made to Ministers asking them to accept the Regulations. They will then be laid before Parliament.

It is anticipated that the Regulations and Guidance will be in place by the required date of 14 May 2000.

Even though the Regulations are in place on 14 May 2000 there will be no immediate effect on product authorisations as the Regulations (and the parent Directive) depend on active substances being listed in annex I of the Directive. On 14 May 2000 Annex I will be empty.

3.5 Transitional arrangements

- BPR DO NOT APPLY UNTIL ACTIVES ARE ON ANNEX I
- ANNEX I IS EMPTY ON 14 MAY 2000
- ANNEX I IS FILLED ESSENTIALLY BY THE REVIEW PROGRAMME

There will be a long transitional period (the Directive foresees 10 years) before the Regulations come fully into effect. During the transition period existing national rules will contine to apply - that means CoPR will continue until all biocidal active substances have been reviewed under the biocides programme.

3.6 The Review Programme

There are approximately 2000 existing active substances to be reviewed over a 10   year period. The European Commission and member states are currently working on producing a review Regulation which is expected to be in two parts. Part one will outline the process and may be accompanied by a list of existing active substances on the market. It is anticipated to be published by the end of this year. Part two of the review Regulation will follow some 18 months later and will contain the first list of active substances to be reviewed.

The review programme is likely to be fairly similar to that operating for agricultural pesticides under Directive 91/414/EEC and its main features are:
- REVIEWS WILL BE CARRIED OUT BY MEMBER STATES
- ANNEX I ENTRY WILL BE DECIDED BY THE COMMUNITY
- THE PROGRAMME IS DUE TO BE COMPLETED IN 10 YEARS

4 EFFECTS ON INDUSTRY

The Directive, and implementing Regulations require that risk assessments are carried out on biocidal active substances and the products containing them. This   requires   the submission, and in many cases the generation of data and industry has to provide this.

4.1 What does industry have to do?

Industry has to consider the following key questions:
- WHAT ACTIVE SUBSTANCES ARE USED?
- WHAT ARE THE DATA GAPS?
- WILL THE DATA GAPS BE FILLED?

Industry has to decide which of the existing active substances it will support and supply the necessary data to the competent authorities. Data must also be submitted on any new active substances (not on the market on 14 May 2000).

Individual companies should try to get together to form task forces to help with   the gathering and submission of data. The Directive encourages this and gives member states the power to require this by law.

Industry should also continue to work with the European commission and other member states on the many working groups which are working on   clarification   of aspects related to the Directive. These groups cover topics such as   the   scope   of   the Directive, the exact data requirements and various European guidance documents but most importantly the review programme.

In summary, even though the Directive itself has been agreed, industry can still influence:
- THE REVIEW PROCESS
- THE DATA REQUIREMENTS
- THE GUIDANCE DOCUMENTS

5 THE TIME SCALE

Assuming that everything goes according to plan and that agreements are quickly   reached the timetable looks like this:
- IMPLEMENTATION IN MEMBER STATES   - May 2000
- FIRST REVIEW REGULATION   - Jan 2000
- GUIDANCE DOCUMENTS   - May 2002
- SECOND REVIEW REGULATION   - July 2002
- REVIEW PROGRAMME STARTS   - Aug. 2002
- FIRST ACTIVE ON ANNEX I   - Aug. 2003
- ALL ACTIVES ON ANNEX I   - Aug. 2012

It has to be said that this is a very optimistic timetable and the author's view is that there is likely to be significant slippage in all phases but especially in the operation of the review programme.

6 SUMMARY

6.1 Positive aspects of the Directive

These are:
- STANDARDISATION OF THE DATA REQUIREMENTS
- MUTUAL RECOGNITION OF PRODUCT AUTHORISATIONS

6.2 Negative aspects of the Directive

These are:
- THE DATA REQUIREMENTS THEMSELVES
- THE AUTHORISATION COSTS
- A RESTRICTED CHOICE OF ACTIVES AS A RESULT

7 CONCLUSION

In conclusion the Biocidal Products Directive will have the following effects:
- NEW REGULATIONS IN PLACE IN THE UK BY 14 MAY 2000
- NO IMMEDIATE EFFECT ON PRODUCT AUTHORISATIONS
- THE REQUIREMENTS COME IN GRADUALLY OVER 10 YEARS

Pulp and Paper

BIOCIDES IN THE PULP & PAPER INDUSTRY: AN OVERVIEW

Dr Pamela Simpson

Whitewater Technologies Ltd, P.O.Box 1153, Stourbridge, West Midlands DY8 2GB

1 INTRODUCTION

Growth of microorganisms in paper manufacturing processes can cause major technical, economic and hygienic problems, mainly due to slime formation. The chemical-physical conditions, and the composition of microorganisms, may vary widely in one process and between processes. Resistant strains can evolve following repeated treatment by certain biocide technologies. Furthermore the regulations for paper intended for food packaging and environmental labelling systems limit the choice of biocide active ingredients.

Biocides are added to the wet end process to prevent slime formation. Introduction of neutral or alkaline sizing instead of acidic papermaking, the closed water circuits, and the increasing proportions of recycled paper have required changes in biocide types in order to control different microbial populations.

Biocides are also used in the pulp and paper industry for the protection of processing materials. The main biocides used in this application fall into the preservative range and can be different from others employed within the paper processing.

This paper aims to introduce the problems that exist within the pulp and paper processing and detail how biocides can help to reduce these problems.

2 THE PULP INDUSTRY

2.1   Types of Pulp

The pulp industry is broken down into predominantly two types of fibre types:

2.1.1 Virgin fibre, which can be further, split into two types, e.g. chemical and mechanical. Chemical virgin fibre undergoes treatments that remove most of the lignin; mechanical undergoes treatment that maintains the level of lignin within the pulp. Both these types of pulp are relatively free from microbiological contamination.

2.1.2 Recycled fibre such as newspaper, coloured paper, computer paper and is usually heavily contaminated with microorganisms.

Both of these types of pulps offer a variety of microbiological problems with the most severe being from the recycled paper types. The microbiological problems encountered are both aerobic and anaerobic bacteria which can develop at different stages of the pulping and can lead to the production of "off" odours usually associated with hydrogen sulphide. Once produced, hydrogen sulphide can react with free metals to produce pigmented sulphides that cause further problems of discolouration.

Wherever virgin fibres are used for pulp manufacture, although these appear relatively free from microbial problems, there may be fungal spores present which originated from the timber itself. If these are presented with the right growth conditions, germination may take effect, causing hyphal development and surface "mat" formation. As a bi-product of this growth, pigments can also be released from the fungal species, which again can cause pigmentation of the pulp.

2.2   Biocides Encountered in the Pulp Industry

Over the last few years, the pulp industry has come under extensive pressure to reduce the levels of halogens being discharged in effluent streams. As the pulping process utilises high levels of halogens during their bleaching process, the level of available organic halogen (AOX) is high and thought to be of concern to the environment. Alternatives have therefore been sought which can achieve the bleaching requirements, but have minimal impact on the environment. These also, indirectly, have had a great effect on the level of microorganisms remaining within bleached pulp. Hence non-sterile pulp may be transferred to a paper process and cause problems during manufacture.

Table 1 below represents the types of biocides that were used during the pulping process and subsequent biocides that have been introduced over the last few years to assist in the reduction of microbial contamination. In general, they types of biocides that are now being utilised are fast acting bactericides to prevent the anaerobic activity which results in the production of odours and blackened pulp.

Table 1: *Biocides used within the pulp industry*

Biocide Type	Status
Chlorine	No longer used
Chlorine Dioxide	Significantly reduced due to AOX
Halogenated Hydantoin	Increase in use due to control release of halogen reducing AOX
Quaternary Compounds	Low foaming types introduced. Not halogenated and fast acting bactericide
THPS	Not halogenated and fast acting bactericide
Glutaraldehyde	Not halogenated and fast acting bactericide

The halogenated hydantoins also form part of the new types of biocides being used. Although these contain halogens, these are released on demand and hence are only available when microbiological activity demands the release of the halogen. Hence the release of halogens is significantly reduced and is seen as a reduced environmental concern.

3 COATINGS AND ADDITIVES FOR PAPER MAKING

3.1 Types of Coatings and Additives

The coatings and additives utilised during paper processing are also a significant source of microbiological contamination. The types that are frequently used are:

> Fillers: e.g. calcium carbonate, kaolin, titanium dioxide
> Sizing agents, starches
> Natural dyes, brighteners
> Retention aids

All these components are from natural sources and are frequently contaminated by aerobic and anaerobic bacteria which utilise these as their food source. The problems encountered are not too dissimilar to those encountered in the pulp industry with colouration being of great concern. Other problems however are due to the loss of viscosity of these products once microbiological digestion is underway. If this occurs during storage in tanks, the whole product is rendered ineffective and cannot be used within the paper process as the microbiological digestion has altered their properties.

Any product that is utilised during paper production will cause paper sheet colouration or loss of desired properties and hence would cause a significant paper quality failure and lost income. It is therefore important to have a good plant hygiene programme in place to monitor carefully the storage products of such products and ensure that the right preservative chemistry is in place to maintain a microbiological free product for long periods of time.

3.2 Biocides Used in the Coatings and Additives Industry

The types of biocides used in this application vary from the rest of the pulp and paper industry. These require to be longer lasting, i.e. sometimes up to 1 years protection is storage tanks and hence fall into the category of preservatives.

Table 2 represents the types of preservatives that are commonly encountered in these products for use in the paper industry.

Table 2: *Preservatives found in the coatings/additives application*

Preservatives Used
Thione
Glutaradehyde
1,2-Benzo-isothiazolin-3-one
Bronopol
1,2-Dibromo-2,4-dicyanobutane
4,5-Chloro-2-methyl-4-isothiazolin-3-one

Chlorine dioxide has also periodically been encountered in these applications in third world countries, but the effectiveness of this biocide would be short lived.

Due to the nature of the paper industry, the paper being produced can be used in numerous different market types; e.g. food contact paper or non-food contact paper. Once the paper industry has become heavily regulated and used these as a guideline for the types of biocides or preservatives that they will allow in their product range. Hence although legislation does not cover coatings and additives in their unused state, their role in the end use influences the type of preservatives that are permitted by the paper maker.

The key legislative requirements are covered by the USA FDA 176.170 (aqueous fatty food contact), FDA 176.180 (dry food contact), FDA 176.230 (Thione use) and in Europe, the German BGVV 36 (food contact). Hence the preservatives used to protect these types of additives need to have these approvals in order to guarantee their safe use in the desired end application.

4 THE PAPER-MAKING PROCESS

4.1 The Process

The paper making process, in very simplistic terms, involves the use of pulp, additives and water to make a final paper sheet. The type of paper being made will influence the levels of all of these components, for example high quality paper will utilise high levels of pulp fibre and additives, whereas tissue paper will use very little. Diagram 1 below illustrates the process.

Diagram 1: *Simplistic illustration of a paper making process.*

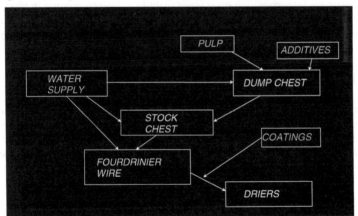

The pulp and paper additives enter the process first through a dump chest in their concentrated form. Adjustments are then made to the concentration in the stock chest just prior to transfer onto the Fourdrinier wire where the paper sheet is produced. Surface additives are sprayed after sheet formation and the final sheet is dried at high temperatures in dryers. The water from the wire is removed into underground tanks and in most cases, recirculated and reused.

As described in earlier sections, there are several potential sources of microbial contamination which can enter the paper mill. Once in the paper processing plant, other factors can then influence the degree of contamination. The conditions within a plant are very often extreme, with high temperatures and humidity arising from the process itself. Hence perfect conditions exist for other microbiological problems to develop.

Contamination sources for micro-organisms can now be extended to the water being reused, air borne species and also man bringing in contamination for external sources. All of these sources are difficult to eliminate due to the nature of the process conditions. Hence the types of microbial species encountered within the paper making process may be more diverse (Table 3), with aerobic and anaerobic bacteria, fungi and yeast appearing throughout the process. In particular, bacterial species such as *Pseudomonas* species are found within the water source and these utilise the food sources available from the additives to produce extra cellular polysaccharides known as slime.

Table 3: *Types of Micro-organisms encountered in the paper making process.*

TYPE	MICROORGANISM	pH
AEROBIC BACTERIA	*Aeromonas*	3.5 - 9.5
	E. coli	
	Pseudomonas	
	Flavobacter	
ANAEROBIC BACTERIA	*Sulfate reducers*	3.5 - 10.0
FUNGI	*Aspergillus*	2.0 – 7.0
	Cladosporium	
	Penicillium	
	Trichoderma	
OTHERS	Yeast/Protozoa/	2.0 - 7.0
	Nematodes	

The development of these biofilms within the system and on the wire can lead to major production problems due to these being transferred onto the paper sheets. Once these are an integral part of the sheet, the drying stage causes shrinkage and holes or tears can occur. This causes considerable down time and loss of income.

In some cases, bacterial spores can also be passed onto the paper sheet and these can withstand the short exposure to high temperature drying. Hence these spores can remain on the paper sheet and if used in food contact applications, could lead to this paper being rejected under quality control procedures.

In some situations, if good plant hygiene is not maintained, fungal spores can also become established in and around the Fourdrinier wire. These can develop into large surface colonies which can become dislodged and transfer onto the paper sheet during formation. Again, this can lead to paper sheet failure.

4.2 Types of Biocides being used in the Paper Making Process

The types of biocides being used in this stage of the process is still heavily regulated by legislation. Slimicides, as they are commonly known, are used to reduce the build up of slime deposits within the water phase of the process. These are again for their safety aspects both to humans and the environment. Hence, like in additives, the numbers of biocides that meet the criteria are very restricted. The USA FDA legislation's not only cover the food contact aspect, but also have a specific approval for the use of slimicides (FDA 176.300). Within Europe, Scandinavia has also introduced their own additional screening programme to protect their environment. This is known as the KEMI approval and the number of approved biocides for paper use has been reduced to approximately 11 actives.

Many changes have occurred over the last 20 years within the European paper making industry. Most systems have now changed from acidic processing to alkaline and the majority have closed up their processing, i.e. the water is now recycled within the plant and not discharged. Hence microbiological contamination have shifted in favour of bacterial strains (due to alkaline conditions) and also increased due to continuous re-introduction of these species and food sources back into the process.

Good plant hygiene procedures are critical to maintaining an effective production, however due to the nature of the process, this will not be wholly effective and the use of biocides is critical to maintain a low microbiological population to prevent paper failure. The table below (Table 4) illustrates the types of biocides that are commonly encountered within the paper processing.

Table 4: *Types of Biocides being used in the paper making process.*

Biocides Used
Halogenated hydantoin
2-Bromo-2-nitropropane-1,3-diol
5-chloro-2-methyl-4-isothiazolin-3-one
2,2-dibromo-2-cyanoacetamide (DBNPA)
n-Octyl-isothiazoloin-3 one
Methylene bisthiocyanate (MBT)
Quaternary ammonium compounds
THPS
Glutaraldehyde

Although these are all used across Europe, the hierarchy of these has altered slightly due to the influences of environmental legislation. For example, Chloro-isothiazolone has been identified as a skin sensitiser and hence therefore lost its number one position as a controlling agent. New actives such as the Octyl – isothiazolone have recently been introduced as a more stable molecule where it is hoped that lower levels can be used reducing environmental and human safety aspects.

In addition to the biocides being used to control micro-organisms within the system, fungicides such as dichlorophen, carbamates, thiabendazole and more recently octyl-isothiazolinone are also used as a surface fungicide. These are applied to specialised papers only such as hygienic paper, mould resistant packaging. Depending on the end-use of this packaging, these may require the food contact approvals.

5 FUTURE IMPLICATIONS

Although there are many biocide alternatives available on the market, for example enzyme technology or bio-dispersants, there appears to be a continued requirement for the use of biocides in order to reduce the levels of microbiological contamination entering the paper making process. The increased awareness of environmental and safety aspects will continue to play an important role on the selection of biocides for paper making processes. The use of legislation to select biocides must be done in parallel with each plants internal risk assessment. No one biocide active will meet all the criteria set out by different European countries and hence the use of these actives must be carefully assessed on a plant by plant basis.

THE APPLICATION OF DIALDEHYDE CHEMISTRY FOR CONTROL OF CATALASE IN THE DEINKING PROCESSES

Per Sundblad, Business Area Manager Deinking
BIM Kemi AB
Box 3102, 443 03 Stenkullen
SWEDEN

Ruarri Scott, Application Engineer
Cellkem OY
Korjalankatu 18, 451 30 Kouvola
FINLAND

1 INTRODUCTION

Increased usage of recycled fiber in combination with system closure and Hydrogen Peroxide (HP) bleaching at moderate temperatures, has developed microbe cultures that are very effective in decomposing (HP). As this problem often comes slowly when the mills are closing their water loops and reuses their wastewater it is sometimes difficult to say when it needs to be treated.

In the beginning, an ordinary biocide might help but normally the mill comes to a point where the consumption of HP has increased to unacceptable levels (15-20 kg/t). Due to varying catalase activity originating from degree of closure, type of water introduced, etc the brightness starts to vary in an uncontrolled way. To compensate for these variations the mills have to control the brightness with trim bleaching (hydrosulphite/dithionite).

After extensive research, followed by numerous full-scale trials, BIM/Cellkem found the way to inhibit/ deactivate catalase and take control over the microbes. During this work we tried almost every biocide one can dream of.

The Cell'link Concept™ was invented and patented in 1994. Since then BIM/Cellkem has developed the invention to a complete concept including a Catalase/Residual peroxide meter from BTG. This is now a turnkey equipment that can be installed anywhere where Catalase needs to be controlled.

Why the need for deinking?
- Abundant source of cheap fibre
- Legislation and regulations
- Environmental issues
- Customer demands

Today's "Sun" can become Tomorrow's "Telegraph"

2 BACKGROUND

The deinking process (DIP) can be compared with a giant washing machine, that removes the printing ink off the fibre, two major types exist:
Wash deinking: often used for flexo-printed waste and in older tissue mills, large water consumption. *Flotation deinking:* for oil-based printings such as offset and rotogravure, technology from the mining industry.

Figure 1

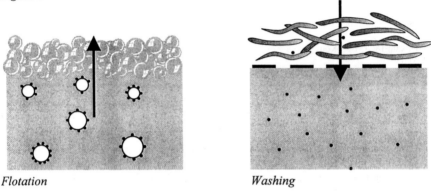

Flotation   *Washing*

Figure 2   *Schematic draw on a two loop flotation deinking process*

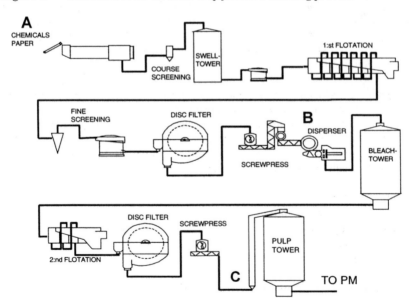

Hence it is the fibres which are valuable. It is of greatest importance to maximize the yield and minimize the water consumption. The process waters become quite dirty, an excellent nursery for bacteria.

- Perfect temperature 45 - 55 °C.
- pH 7 - 9
- Unlimited food supply (hemi-cellulose chains and other nutrients).
- Aerobic conditions topped up with some HP.
- Generation time of about 20 minutes.
- Unlimited number of sites to sit on.
- Unlimited numbers of "new friends" entering the area with raw materials and waters.
- The "by products" being automatically transported away.

During these perfect conditions the reproduction of microbes will be almost unlimited. As HP is excellent for environmentally friendly bleaching these microbes will be teased with HP to produce as much Catalase as possible. This will soon lead to unacceptable HP consumption.

Catalase is one of the oldest known enzymes and was developed when "life" became aerobic, i.e. when organisms started to use oxygen. When a cell uses oxygen in its metabolism HP is very often produced as a by-product. However HP is toxic to the cells, so they need some defence mechanism. Catalase was "invented" by the evolution to protect living cells from HP.

Catalase is by far one of the fastest enzymes that exist. One molecule of catalase can easily decompose 1 million HP molecules per second or one kilo catalase can decompose 300 kg 50% HP per second.

What we need is a broad spectra "medicine" that will react permanently with Catalase and Peroxidase without giving any harmful or toxic residual-products.
It has to be readily biodegradable and co-operative with HP

3 CHEMISTRY AND BIOLOGY

Cell'link™ is based on the chemistry of dialdehydes and their cross-linking reaction with amino-groups of the protein chains.

Figure 3   *Dialdehyde reaction with the amino groups of the protein chains*

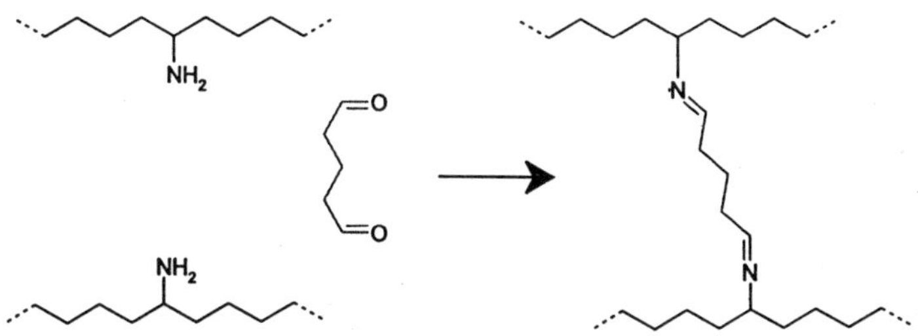

Then when amino groups are cross-linked on the right spots, the enzyme is blocked out from further HP decomposition reactions. Of course, the dialdehydes are not specific either to the enzyme or to the bacteria, but there are enough dialdehyde molecules present to decrease the enzymatic activity to an acceptable level.

So far there is no known defence mechanism against dialdehydes in living cells and the risk for adaptation should therefore be minimal.

The enzyme is produced by aerobic microbes, which live in the mill waters and in the bio-film located on all wet surfaces. When these bacteria are teased with low concentrations of HP, which is the case in all mills that are using HP, the population will change so that the individuals with the highest catalase activity will have the best opportunities to survive. This adapted population grows and infiltrates the whole circulation water system.

Catalase and Peroxidase are produced inside the microbes and will be located in the cell wall. When these cells pass away either due to age or due to external circumstances their content will spread out in the water. These free-swimming molecules of Catalase are the worst due to that HP have the best accessibility to them, *the reaction could not be faster.*

3.1 How to get control over the microbes instead of letting them mess up your process?

- Something that inhibits the free enzyme.
- Something that can keep the microbes on an acceptable and controllable level.
- Something that inhibits the newly produced Catalase.

This product has to be:
- Cost effective
- Readily biodegradable
- Free of toxic residual products

Since more microbes are introduced all the time into the loop we need to control the amount of catalase continuously. One of the few additives that can do all this is the **Cell'link Concept**™ in synergy with HP.

BTG has invented a Residual HP and Catalase activity detector, which has been further developed in co-operation with BIM to a complete control-system for DIP water systems. Now it is possible to balance on that rope to keep the RHP with a minimum of added Cell'link and still have a complete control over the microbes. This system will detect any variations in incoming water or buffer water and it will act according to that.

Figure 4   *BTG Catalase activity and Residual Peroxide Meter*

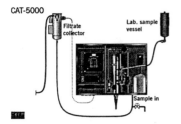

3.2 Mill experiences

The **Cell'link Concept** ™ has successfully been running in more than 10 DIP mills over the past years. **Not a single mill** has so far switched to something else. Even if numerous other trials have been done we have not yet heard of any other real success.

Case study: DIP mill in Northern Europe produces Newsprint from DIP on two lines: 70% ONP, 30% OMG. Approximately 300 t/d in each line. Prior to the trial the mill was suffering from severe decomposition of HP and was unable to control the brightness properly.

The Key targets were to get:
- Increased and controllable brightness.
- Increased yield.
- More uniform pulp quality.
- Decreased bacteria level in Loop 1.
- Residual peroxide throughout whole Loop 1.
- Saving of chemicals.
- Improved process balance.
- No disturbance in other parts of the process.

The pre-trial survey showed:
- $\approx$ 75 % of the HP was decomposed within minutes after dosing.
- Large variations in pulp quality.
- No residual peroxide.
- High bacteria levels ($>10^8$ μ/ml) of which almost all were catalase positive.
- How and where the different waters were flowing.
- Where the water dividers were.
- Water sources.
- Type and shape of the raw material.
- Equipment and layout

With this in our hand we were able to define our plan of attack.
- Dose point
- Dose rates
- Dose strategy

Cell'link ™ was dosed to Loop 1 water and after eight hours it was possible to detect residual HP in water going into the common filtrate tank. After two days residual HP was detected even in the water going into the pulper.

All the Key targets that we had set up together with the mill were achieved within a couple of days and in addition the mill could also see:
- Decreased usage of Hydrosulphite (HS) in the trim bleaching.
- The amount of broke due to unwanted brightness drop was significantly reduced.
- Bacteria levels decreased to $\approx 10^4$ μ/ml (a decrease of more than 99.99 %).
- The old biocide treatment could be terminated.

Later research and full-scale trial have shown that bacteria levels higher than 10^6 μ/ml are dangerous and it is just a matter of time before problems return. If the levels on the other hand are lower than 10^4 μ/ml the process will be stable and it is unlikely that disturbances will occur. (The absolute levels of bacteria may vary due to catalase activity and method of measurement).

Figure 5   *Chemical addition on Line 2*

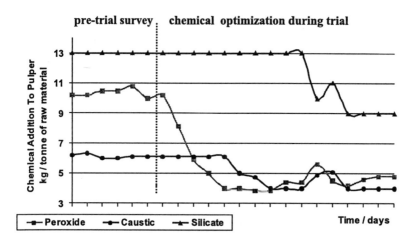

Figure 6   *Chemical addition on Line 1*

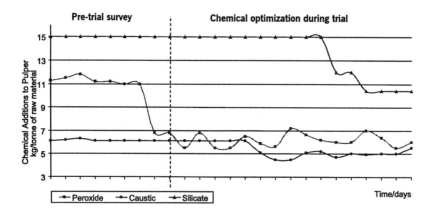

Figure 7 *Decomposition rates during a startup with the Cell'link Concept*

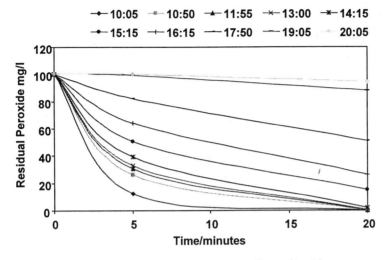

Figure 8 *Decomposition rates using an ordinary biocide*

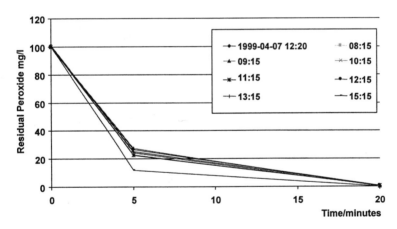

Table 1 Potential savings for an ordinary size DIP mill

	Kg/t	Eur/t	£/t
HP	8	3.8	2.58
Caustic	2	0.36	0.25
Silicate	2	0.43	0.30
Trim Bleach	2	1.6	1.09
Broke*	1%	3.7	2.52
Total		9.9	6.74
Cell'link Cost	0.3	1.5	1.00
Net / ton		8.4	5.74

Calculated on a production of 2 x 300 t/d as for this case study:
- 1.7 Million Eur/year
- 1.17 Million £ / year

4 CONCLUSIONS

- Of all mills that have tried **Cell'link**™ not one has quit using it.
- No detected by- or side effects of Cell'link usage.
- Bacteria levels decreased by a factor 10,000 or more.
- No ordinary biocide has to be used.
- **Cell'link** ™ dose rates from 0.1 to 0.4 kg/t.
- No known risk that the microbes could develop a defence against Cell'link.
- Very significant savings of chemical costs.
- More stable and reliable operations.
- Increased balance in the process chemistry.

Water

THE NOVEL USE OF CHLORINE DIOXIDE IN THE TREATMENT OF DOMESTIC WATER IN BUILDING SERVICES

A Harris and T J Parkinson

Feedwater Ltd.
Moreton, Wirral
L46 4TP, UK

1 INTRODUCTION

The potential health hazards associated with the presence of Legionella Pneumophila in water systems where aerosols can be created have been well documented[1,2].

There are now in the UK regulations and guidelines which prescribe the practice to be adopted to minimise the risk of outbreak of disease associated particularly with Legionella Pneumophila[3].

The hot and cold water services in buildings have been identified as potential sources of Legionella proliferation. Whilst the risk of Legionella proliferation in such systems may be minimised by good system design, in practice it is found that many systems do not conform to the ideal and their redesign would be economically prohibitive..

The use of on-line treatments has therefore been adopted in many instances as the means of minimising the risk of Legionella proliferation. The most commonly used treatment is to inject chlorine (usually in the form of hypochlorite) into the system to maintain a reserve of 2ppm free chlorine. In many cases this proves to be effective. However the use of chlorine raises a number of concerns.

The use of chlorine at such relatively high concentrations may give rise to the production of chlorinated by-products if the water supply contains traces of chemicals which can react with chlorine. Such trace chlorinated chemicals are known to be carcinogenic[4] and their presence in water supplies for <u>possible</u> human consumption should be minimised.

Biofilms which may develop in water systems are known to provide ideal habitats for the survival and growth of Legionella[5]. Chlorine does not readily penetrate such biofilms and destroy Legionella[6].

The effectiveness of chlorine reduces markedly as the pH of the system water rises above 7 and at pH's above 8 is largely ineffective as a disinfectant. High levels of chlorine are also known to be aggressive to many of the materials of construction of water services.

There is thus a requirement for a disinfection system which will assist in minimising the risk from Legionella in hot and cold water services without the problems associated with chlorination.

2 CHLORINE DIOXIDE

Chlorine dioxide has been used widely in Europe since the early 1940's as a drinking water disinfectant. More recently the USA has suggested the use of chlorine dioxide to reduce the formation of chloro-organic compounds particularly chloroform and other trihalomethanes (THM's) which are known carcinogens[7].

The use of chlorine dioxide in water systems results in its reduction to chlorite and chloride. In the UK the Drinking Water Inspectorate (DWI) restricts the use of chlorine dioxide in potable water supplies to a maximum of 0.5ppm total oxidants expressed as chlorine dioxide. This ensures that chlorite (and any chlorate) concentrations do not reach levels of potential harm to humans.

Chlorine dioxide is an effective oxidising agent with broad spectrum biological activity which does not take part in substitution or addition reactions with organic chemicals. As a gas it cannot be compressed or stored in any concentrated form, so that all applications of the chemical have required production at the point of use. Production of chlorine dioxide for water treatment applications usually utilises sodium chlorite ($NaClO_2$) solution as the starting material which is then reacted either with excess chlorine gas or sodium hypochlorite and acid or acid alone. These processes all require equipment and systems to ensure correct reaction stoichoimetry and safe storage of the concentrated chlorine dioxide solutions produced. The potentially hazardous nature of the gas has, in the past, normally required a carefully designed plant and skilled operators to minimise potential hazards, and this with the high capital investment required, and the difficulty in controlling the concentration, has precluded its use in 'low volume' applications such as building services.

Solutions of 'stabilised' chlorine dioxide have been claimed to provide a simpler system for delivering chlorine dioxide. However the effectiveness of such solutions depends on the amount of chlorine dioxide released, which depends on the strength of acid used and length of time for reaction[8].

The ACTIV-OX® system has been developed to meet the needs for a 'safe' and controllable chlorine dioxide system for application in small water using systems. The system instantaneously delivers over 90% of the available chlorine dioxide at a pH of 4 compared to other systems which require lower pH and or longer reaction times (Fig 1 and 2).

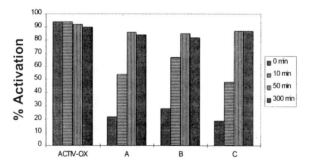

Fig 1 Comparison of Chlorine Dioxide Pre-cursor
Activation with pH   (after 5 min)

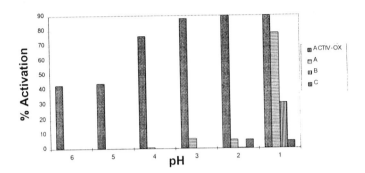

Fig 2   Comparison of Chlorine Dioxide   Pre-cursor
Activation with time

The ACTIV-OX® chlorine dioxide system evaluated in this trial overcomes many of the problems associated with chlorine dioxide for the "small" water user. A chlorine dioxide precursor solution and a dilute acid solution are mixed in a 1:1 ratio immediately prior to injection into the water to be treated. The dose rate of chlorine dioxide is controlled by water meter signal to two proportioning pumps. The mixing of the two chemicals immediately produces a chlorine dioxide solution which is diluted to the required strength by injection into the water to be treated (Fig 3).

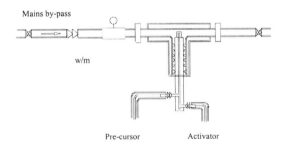

Fig 3 Chlorine Dioxide Injection system

The ACTIV-OX® system minimises potential hazards as there is no storage of concentrated chlorine dioxide solutions and failure of either of the injection pumps or loss of water flow results in no chlorine dioxide production.

This system enables the 'low volume' water user to accurately and safely utilise chlorine dioxide without the need for expensive capital equipment or strong acid solutions.

3 THE HOT AND COLD WATER TRIAL SYSTEM

The system comprises a cold water storage tank of 0.5m³ capacity which feeds a steam heated calorifier and then to a series of warm water mixer taps and showers as shown schematically in Fig 4.

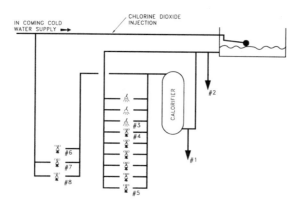

Fig 4 Schematic Diagram of Hot and Cold Water system

The system has limited usage and no hot water recirculating pump - in effect it presents an ideal environment for the growth of Legionella and the worst features of system design:

- long residence time cold water tank exposed to solar radiation
- long residence time calorifier with no recirculation
- hot and cold water pipes alongside each other
- little used taps at end of pipe runs

Prior to the injection of any treatment the microbial population was assessed for six weeks. This showed considerable Legionella populations at all outlets including a small amount of activity at the hot water tap (#6) even when the calorifier was operating at a temperature greater than 60°C.

From week 7 chlorine dioxide was injected into the inlet water to the cold water tank.

The assessment of the performance of the new chlorine dioxide system was carried out over a continuous period of 39 weeks.

4 RESULTS AND DISCUSSION

The concentration of chlorine dioxide, chlorite and total oxidants was determined on site using a portable colorimeter (Palintest Photometer 5000) and a modification of the DPD test in which any chlorine species are complexed with glycine to ensure only chlorine dioxide reacts with DPD. The chlorite and total oxidants are then determined on a fresh sample by acidification and neutralisation in the presence of potassium iodide. The initial dose level was set at 0.3ppm chlorine dioxide injected in the water feed to the cold water tank. This was increased to a maximum of 1.0 ppm and then reduced to 0.5ppm. Relatively quickly a reserve of chlorine dioxide was established in the tank and the bottom of the calorifier.

Levels of Legionella were determined weekly using a method derived from British Standard Draft Directive 211 (1992) using modified Wadowsky Yee medium

Samples were cultured for 10 days and results expressed as colony forming units per litre (cfu/l)

The levels of Legionella at points in the system are shown in Figs. 5,6,7,8. It can be seen that control of Legionella in the feed tank and calorifier bottom was quickly established once chlorine dioxide injection commenced at week 7. (Figs 5 and 6).

The control of Legionella at other outlets of the system took much longer to establish(Figs 7,8).

The presence of biofilm is now accepted [5] to provide an environment where Legionella can develop and that destruction of biofilm is essential to the establishment of a Legionella free environment.

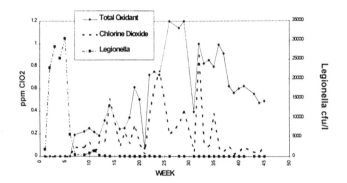

Fig 5 Field Trial Sample Point #2

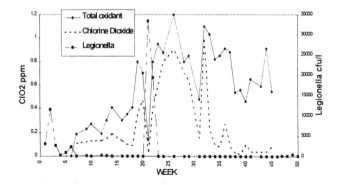

Fig 6 Field Trial Sample Point #1

The usual method of sterilising a system is to circulate high levels of disinfectant (usually 50ppm Cl_2) for periods of at least one hour[3]. During this trial there was no attempt to clean the system in this way.

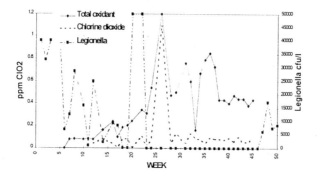

Fig 7 Field Trial Sample Point #5

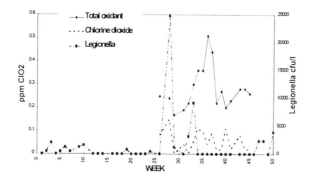

Fig 8 Field Trial Sample Point #8

The effectiveness of chlorine dioxide will depend not only on the concentration of the chemical but also on volume of water passing through a system. The effect of system volume is often ignored when considering the effectiveness of treatment but it is clear that as a certain <u>weight</u> of an oxidising biocide will only be able to kill a certain number of bacteria then the more <u>weight</u> of biocide added the greater will be the effect (i.e. the greater the water used at a fixed biocide concentration).

The system under consideration had a very low water usage, on average about $6m^3$/week, thus the <u>weight</u> of chlorine dioxide used was only 3gms/week (0.5 x 6). It is not unexpected therefore that the cold water tank and the calorifiers were the first parts of the system to be free of Legionella because, due to their large water volumes, the amount (weight) of chlorine dioxide was greatest in these areas.

To control Legionella throughout the system two strategies were adopted.

- the dose level of ClO_2 into the feed water was increased to 1.0ppm
- automatic timer valves were installed at two points in the system (Fig 4) to ensure that some water flowed each day through the system.

The effect of these changes was to ensure that some chlorine dioxide was able to reach the extremes of the system. The system showed no detectable Legionella at all outlets except #8 by week 24.

Low levels of Legionella continued to be detected at outlet #8 - a mixer tap at the extreme of one leg of the system. A closer investigation determined that this tap was fed with cold water from a mains supply and not the treated cold water tank. The already low concentration of chlorine dioxide at this end of the system was thus being reduced further by mixing with an untreated supply. The tap was dismantled, swabs from the internals showed positive Legionella on the 'fabric' washer. Disinfection of the fitting in a 50ppm (Cl_2) solution of sodium hypochlorite was carried out and no further Legionella were detected. (Fig 8).

The calorifier temperature at the commencement of the trial was 60-63°C providing an outlet temperature of 55°C at the single non-mixed tap of the system. Whilst no Legionella were detected here there was little evidence of any chlorine dioxide or total oxidants either. The effect of this high water temperature and long calorifier retention time was to cause the loss of most of the chlorine dioxide through the calorifier vent due to increasing volatility at elevated temperatures. In this system this high temperature and low ClO_2 had little consequence because the temperature regulated valves ensured that a high proportion of treated cold water mixed with the hot water ensuring that chlorine dioxide was present at the outlets (except in the case of #8 as already described). The temperature of the calorifier was progressively reduced from week 37 until at the end of the trial it was 45°C without any adverse effect on the freedom from Legionella. A further study[9] has also shown the effectiveness of chlorine dioxide in hot water systems at reduced temperatures. In a large system the ability to reduce calorifier temperatures whilst still maintaining Legionella free conditions would obviously have a significant financial benefit.

5 CONCLUSIONS

The newly developed ACTIV-OX chlorine dioxide system has effectively delivered continuous low levels of chlorine dioxide which were effective in controlling Legionella in a hot and cold water system without the need for prior disinfection.

Careful attention needs to be paid to system design if the application of chlorine dioxide is to be successful.

- low flow areas of the system should be minimised
- recirculation should be introduced where possible
- reduction of calorifier temperature should be considered to increase oxidant levels at outlets

References

1)   G. C. Clifford, "Handbook of Chlorination and Alternative Disinfectants" , p.314
      Van Nostrand Reinhold, New York (1992)

2)   Puckorius, P.R. "1995 Legionnaires Disease Outbreaks in USA Cooling
      Systems" paper presented at 1st Australian Conference of the International
      Ozone Association February 1996

3)   "The control of Legionellosis including Legionnaires' Disease" HS(G)70 HSE
      Books ISBN 0 7176 0451 9

4)   Noack, M.G. and Doerr, R.L. Water Chlorination 2 49-58 (1978)

5)   Rogers, J. *et al* Applied and Environmental Microbiology 60 1842-51 (1994)

6)   Walker, J.T. *et al* Biofouling 8 47-54 (1993)

7)   Colclough, C.A. *et al* Water Chlorination 4 219-229 (1983)

8)   Kirk-Othmer "Encyclopaedia of Chemical Technology" Vol 5 p.991 (1993)

9)   Pavey, N.L Roper, M "Chlorine dioxide Water Treatment - for hot and cold
      water services"
      Technical note TN2/98 BSRIA. ISBN 0 86022 486 4

BIOFILMS: ADVERSE ECONOMIC IMPACTS AND THEIR AVOIDANCE

A. J. McBain and P. Gilbert

School of Pharmacy and Pharmaceutical Sciences
University of Manchester
Manchester M13 9PL

1 INTRODUCTION

In the majority of situations where micro-organisms grow and survive, they can be found in close association with surfaces, as sessile communities or biofilms. Biofilms are functional consortia of cells bound within exopolymer matrices and organised at interfaces. Although the unique physiological properties of microbial biofilms are now recognised as important in many aspects of life and commerce, biofilms are more usually associated with problems. These relate not only to oral disease and dental caries, but also to infection, public health and commerce. For example, biofilms are implicated in infections related to medical implants, are associated with cross-contamination in food manufacturing, with problems in domicilliary hygiene, in the biofouling of pipes, and in biodeterioration. In these respects a major concern of biofilm research relates to biofilm prevention and control. The nature of commercial screens for lead molecules means that antibacterial agents have often been developed and optimised for their activity against fast-growing, dispersed, planktonic populations of individual species. Such agents are generally ineffective when deployed against microbial communities growing as biofilms which are reported as being some 100-1000 times less susceptible than their planktonic counterparts.

Recalcitrance of biofilm communities to antimicrobials has been attributed to a variety of mechanisms. Briefly, the close juxtaposition of cells within a glycocalyx, composed of extracellular polymers and incorporating entrapped extra-cellular enzymes, establishes a reaction-diffusion barrier and leads to poor penetration of oxidising biocides, antibiotics and nutrients. As a result, deep-lying cells within biofilms are not only exposed to lower levels of antimicrobial but they are also severely nutrient and oxygen-limited. The latter causes the expression of starvation phenotypes including multi-drug efflux pumps and enhanced exopolymer synthesis. During exposure to anti-microbial agents these slower-growing organisms are exposed to sub-lethal levels of agent, and will generally out-survive their less nutrient-depleted congeners. Regrowth of the population will enrich for drug resistant phenotypes and genotypes during the post-treatment phase. This article will consider biofilms in the context of economic importance and review the current understanding of biofilm drug-resistance mechanisms, together with sophisticated, novel approaches for biofilm control, prevention and eradication.

1.1 Microbial Biofilms in Context

Biofilms are heterogeneous populations, which encompass bacteria, fungi, algae and protozoa. They are ubiquitous in nature[1,2] and represent populations of cells that have functional interdependencies upon one another which collectively offer microbial activities that are not possible by any of the individual component species.[3] Biofilm-associated microorganisms play crucial roles in terrestrial and benthic nutrient cycling[4,5] and in the biodegradation of environmental pollutants.[6,7] Bacteria that are associated with mammalian skin and mucosal surfaces protect the host from pathogenic bacteria whist sessile communities in the human large intestine play an important role in the metabolic well-being of the animal.[8] In natural water systems, surface-associated microorganisms degrade organic compounds and detoxify xenobiotics, thereby maintaining some aspects of water quality. Furthermore, the metabolic activities of microbial biofilms have been harnessed for wastewater management,[9] sewage treatment[10] and in biotechnology, in a variety of solid-state fermentation processes.[11]

The downside of microbial biofilms is associated with their involvement in major problems associated with industry, medicine and everyday life. In industry, biofilms are responsible for significant losses of efficiency, process down-time and damage of equipment, together with biofouling of pipelines, process equipment and heat exchangers.[12,13] During food processing, product contamination can occur as a result of adherent-biofilms on food-contact surfaces,[14,15] whilst considerable energy dissipation occurs in marine transport due to biofilms formed on the hulls of ships. In medicine biofilms are now firmly associated with a majority of chronic infection scenarios.[2] The economic implication is the loss of billions of dollars yearly, world wide

Physical methods for the control of microbial biofilms, although often effective, are in many situations impractical. In this context it is notable that an almost universal feature of the biofilm mode of growth is their profound resistance to antibacterial compounds. Conventional chemical control methods, developed for use against fast-growing planktonic cultures are only poorly effective against biofilm bacteria. Large doses of biocide or antibiotics, which are either environmentally undesirable or above toxic thresholds respectively, are required to eradicate biofilms in industry and medicine.

2 MECHANISMS OF BIOFILM RESISTANCE

In order to understand current approaches for prevention and control of biofilms, we must first consider the reasons for the failure of conventional antimicrobial protocols. There are thought to be three main reasons as to why biofilm bacteria out-survive their planktonic counterparts during antimicrobial treatments (reviewed by McBain *et al.*[16]).These are: i) poor penetration of antimicrobial compounds due to the presence and turn-over of exopolymer slime (glycocalyx); ii) the imposition of extreme nutrient limitation within the depths of the biofilm community and the co-incident expression of metabolically-dormant, recalcitrant phenotypes; and (iii) the expression of attachment-specific phenotypes that are radically different and intrinsically less susceptible than unattached ones.

An intuitive explanation of biofilm drug resistance is that antimicrobial compounds are physically excluded from the community by the barrier properties of the glycocalyx. Such intuition however envisages that the glycocalyx functions as a biocide-impermeable umbrella, but since it generally possesses a diffusivity approximating that of water, this alone does not present an efficient barrier to penetration.[2,17-19] Poor drug penetration into the biofilm must therefore involve some inactivation or removal of the agent by the glycoclayx, *viz* reaction-diffusion limitation. Such inactivation of biocide might involve simple ionic interaction with the glycocalyx acting as an ion-exchange resin or covalent interaction of reactive compounds such as the oxidising biocides iodine, iodine-polyvinylpyrolidone complexes,[20] chlorine and peroxygens[21] with the glycocalyx. Alternatively, enzymatic inactivation of certain less reactive or uncharged antimicrobial molecules might augment the barrier properties of the matrix. With respect to enzyme-mediated reaction-diffusion limitation, enzymes such as ß-lactamases,[22] formaldehyde lyase and formaldehyde dehydrogenase[23] become concentrated within the glycocalyx and will neutralise susceptible antimicrobials as they diffuse into the community. In these instances, enzymatic hydrolysis of the antimicrobial can lead to severe penetration failure, provided that the turn-over of enzyme is sufficiently rapid.[19]

Whilst reaction-diffusion limitation may help to explain resistance during brief exposures to biocides, such as in single applications, or where the exposure is transient (i.e. in clean in place), it cannot account for survival during long term treatments. In such situations the biocide present within the bulk phase will augment that quenched by the biofilm, and will eventually saturate all possible reaction / binding sites. Resistance can only then be obtained when the bulk phase has become fully depleted of antimicrobial agent. Thus reaction diffusion limitation should not be thought of as an impenetrable umbrella but as one that is simply shower-proofed. Clearly other mechanisms of resistance associated with biofilms must compound that of diffusion limitation.

In this respect, an important resistance mechanism is associated with the variety of growth rates and nutrient-deficient phenotypes present within the community, and relating to nutritional and gaseous gradients throughout the biofilm community. It has long been recognised that the susceptibility of bacterial cells towards biocides and antibiotics reflects their physiological status in relation to their growth environment. Within a biofilm community a plethora of phenotypes are expressed at any particular time and the survival of the community then reflects the susceptibility of the least susceptible one[24]. Gradients of nutrient availability ensures that in the depths of a biofilm, growth rates, and hence susceptibility, are generally suppressed relative to planktonic, or outlying biofilm bacteria.[25,26]

Gilbert *et al.*[27] demonstrated this facet of biofilm resistance using perfused biofilm fermenters to control growth rate in attached populations of bacteria and conventional continuous culture to provide appropriate planktonic controls. In this fashion the separate contributions towards resistance, of growth rate and the expression of biofilm-specific phenotypes were evaluated without the imposition of a biocide concentration gradient. A series of publications[25-27] demonstrated that much of the resistance of both Gram-positive and Gram-negative biofilm communities towards a wide variety of antibiotics and biocides is attributable to the existence of physiological gradients of growth rate and the presence of a variety of nutrient-depleted phenotypes (recently reviewed by Gilbert and Allison[28]).

Although reaction-diffusion limitation and the presence of nutritionally restricted phenotypes are obviously important determinants of biofilm drug resistance, neither, either separately or in combination, provides a complete explanation of the phenomena. Cells on the periphery of the biofilm, subject to nutrient fluxes similar to planktonic organisms would succumb to antibacterial concentrations that are effective against the planktonic cells. Cell-death at the periphery would lead to increased nutrient availability for deeper-lying cells. These would, in turn, grow faster and adopt a more susceptible phenotype. The end-result would be a progressive destruction of the biofilm as visualised in many confocal micrographs.[21] Whilst such processes would clearly delay killing within the depths of a thick biofilm community, they do not explain resistance to a sustained antimicrobial treatments. Further, additional mechanisms must be associated with the realities of biofilm-recalcitrance.

The existence of specific, biofilm-associated, drug-resistant phenotypes has been demonstrated by Ashby *et al.*[29] and Das *et al.*[30] for antibiotics and biocides respectively. They showed that initial attachment to a surface could rapidly confer 2-4 -fold decreases in susceptibility towards a broad range of agents. Davies *et al.*[31] demonstrated that such resistance might be mediated by homoserine lactone communication signals actively deployed within the attached community, causing the expression of a biofilm-phenotype. As for reaction-diffusion limitation and nutrient depletion, these changes are insufficient, in their own right, to account for the degree of resistance reported in biofilm communities. This has led researchers to consider gradual selection of resistant phenotypes and genotypes within sub-lethally treated biofilms as a plausible mechanism for sustained, high-level, biofilm resistance.

Consumption of antimicrobials and nutrients by the biofilm provides for chemical gradients with respect to biofilm depth. As a result, certain deeper-lying community members are not only expressing less susceptible phenotypes but are also exposed to sub-lethal levels of antimicrobial over a prolonged time scale. These organisms and their ecological partners will therefore be subject to selection pressures both for increased drug-resistance and for general survival within these highly competitive micro-niches. Such exposure might cause the induction of or enrichment for multi-drug efflux pumps or particular clones or species within the community. Following removal of the stress, either by the removal of the agent or by the death of the out-lying cells, the surviving, less-susceptible phenotypes will come into dominance.

Exposure of cells to sub-effective concentrations of antibiotics, such as tetracycline and chloramphenicol, and to xenobiotics, such as salicylate, chlorinated phenols etc.[32,33] up-regulates the expression of multi-drug resistance operons (i.e. *mar)* and efflux pumps such as *acrAB.*[34] Maira-Litran *et al.*[35] demonstrated that constitutive expression of *acrAB* can protect biofilms against low concentrations of ciprofloxacin and that the expression of *mar* and its target genes is inversely related to specific growth rate. Hence, following exposure of biofilms to sub-lethal levels of ß-lactams, tetracyclines, salicylates or other inducer substances, *mar* expression will be at its greatest within the depths of the biofilm, where growth rates are most suppressed. This will provide an additional survival advantage to these cells. Other multidrug efflux pumps, under the regulation of different inducing agents might extend this explanation of biofilm tolerance to include other treatment agents.

The resistance of biofilm communities to treatment with chemical and antibiotic agents is therefore highly complex and multifactorial. Improvements in our understanding of the physiology and ecology of biofilm communities has enabled better, more specific anti-biofilm compounds to be developed, such as those that target biosynthesis of the glycocalyx and the adoption of a biofilm phenotype. There has also been considerable interest in developing intrinsically biofilm-resistant materials for use in industry and medicine where the financial implications of biofilm development are most severe. This article will go on to outline some of the major new strategies for biofilm control, some of which promise to lessen their considerable economic burden.

3 METHODS FOR THE CONTROL OR ERADICATION OF BIOFILM

In as much as biofilm communities are both problematic and resistant to conventional disinfection / treatment strategies there has been considerable research activity intended to develop strategies that prevent both the initial attachment of cells to surfaces or arrest biofilm development at an early stage of their formation. Since a major contributor to the resistance of treated biofilm is the inability of agents to effectively penetrate the exopolymer matrix and to kill the deeper-lying, slow-growing cells, a number of approaches to the control of biofilms have involved the delivery of treatment agent from the colonised surface. Alternative approaches are to manufacture surfaces, or surface coatings, that have intrinsic bactericidal properties, whose physical nature discourages the formation of the biofilm, or which release anti-signalling molecules that interfere with induction of the biofilm phenotype.

3.1 Antimicrobial-Impregnated Substrata

A wide variety of materials have been impregnated with antimicrobial compounds or coated with biocide in order to create inherently colonisation-resistant materials. Such approaches are unfortunately largely unsuccessful when evaluated for use in industrial applications, since they provide for reservoirs of agent that eventually become depleted. Whilst a slow release of biocide provides for a local high concentration of agent it is also rapidly disseminated to more remote locations where reduced concentrations might also provide selection pressures towards less-resistant bacteria.[36] Additionally, unless the surfaces are also self-cleansing, the attached, killed cells will accumulate and generate thick conditioning films that eventually overcome any residual antimicrobial effect. This conditioning film may then itself provide a substrate for colonisation and growth.

In the above strategies, the rate and extent of leaching of the antimicrobial from the surface is related to its water-solubility and bulk-phase concentration. Organo-silver complexes and silver ions are highly active biocides and represent very low water solubility products. Accordingly such molecules have been widely evaluated in these applications. Rogers *et al.*[37] for example, studied the resistance towards colonisation of silver-coated surfaces by a consortium of organisms that included *Legionella pneumophila*. Although the initial rate of colonisation was retarded, biofilm formation still progressed to completion. It appeared that pioneering species, such as *Methylobacterium* and *Pseudomonas*, had formed a protective layer of cells on the surface to which more susceptible species could then establish. It seems likely in non-enclosed industrial applications, or for instance on the hulls of ships, where there is a constant challenge with micro-organisms that simple biocidal surfaces such as these are unlikely to prove effective. Biomedical applications of anti-biofilm coatings do, however, present a very different situation. Whilst venous and urinary catheters may be under constant risk of bacterial colonisation, indwelling medical devices are only at risk from micro-organisms during and immediately after their implantation. Implant-related infections therefore generally involve mono-species biofilm[2]. In the former category of use, the performance of silver-coated urethral catheters has been disappointing. Riley *et al.*[38] failed to demonstrate any efficacy for such catheters over conventional ones in the incidence of catheter-associated urinary tract infection. Stickler *et al.*[39] reported that silicone urinary catheters, impregnated with ciprofloxacin failed to resist colonisation by sensitive strains of *Proteus mirabilis*, *Pseudomonas aeruginosa*, *Escherichia coli* and *Providencia stuartii* over 48 h exposure periods. Similarly, vascular catheters impregnated with silver sulphadiazine and chlorhexidine completely lost their antibacterial activity after 10 days of use.[40] Based on current evidence, simple antimicrobial surfaces are unlikely to be particularly useful in such applications.

Medical implants such as shunts or artificial joints, on the other hand, only present a minimal opportunity for infection which occurs during or immediately after their implantation. This will minimise the opportunity for bacterial surface occlusion and antibacterial leaching, and also limits the time-span over which the protective capacity of the coating will be required. Accordingly, Bayston[41] has suggested that impregnation with the appropriate antibiotics may prove efficacious in such applications. Silicone shunts were impregnated with a range of antibiotics, and their ability to resist bacterial colonisation was assessed *in vitro*. The shunts were challenged with single doses of *ca.* 10^7 staphylococci and coryneforms and were perfused with liquid culture medium for 14 days before examination. Whilst trimethoprim, clindamycin, spiromycin, and sodium fusidate-impregnated catheters did not resist colonisation over this time period, those treated with rifampicin or combinations of rifampicin with either trimethoprim or clindamycin did. The catheters were re-challenged and by day 28, only clindamycin and rifampicin combinations had proven their continued effectiveness. The shunts treated with these antibiotics also went on to resist a third challenge. The authors suggested that this degree of resistance was sufficient to eliminate nosocomial infections associated with the use of such implants.[42]

From the foregoing, it seems likely that apart from a small number of specialist medical applications, the efficacy of surface coated devices may be compromised by antibiotic-resistant bacteria, together with the barrier effect provided by conditioning films that will rapidly coat biomaterials *in situ*.[43]

3.2 Erodable Biocide-Containing Surface Coatings

In order to overcome the problems of surface occlusion and antimicrobial leaching, there is considerable interest in the development of self-cleaning surfaces that ablate during exposure to fluid dynamic forces. Such ablation might not only remove residual attached cells but it might also serve to release further biocide. In this fashion Cooksey and Wigglesworth-Cooksey[44] have reported some success in preventing biofouling of marine surfaces using erodable, biocide-containing surface coatings. These researchers incorporated biocide into a soluble matrix-coat that enabled dissolution, and hence release, of biocide into the base of the biofilm. In this instance the extent of biocide dissolution depended upon fluid dynamic forces directed at the coat matrix.

Self-polishing coatings, whose release of biocide is independent of fluid dynamics, include several organo-tin acrylates. Such polymers will slowly hydrolyse in water, and will release the incorporated biocide continuously according to the rate of hydrolysis.[45] Suzanger *et al.*[46] evaluated a novel biocide-containing coating material (patent WO 97/05182; PCT/GB96/01617) for which hydrolysis facilitates the release of biocide and where the release of biocide renders the materials prone to surface erosion. The efficacy of a number of such novel polymer coatings, containing various levels of three quaternary ammonium biocides proved disappointing. Ablation and biocide-release appeared to be non-specific and was very rapid over the first few days of submersion in water. Complete loss had occurred after *ca.* 5 days. It was concluded that whilst the approach had much potential as a means of overcoming occlusion caused by attached cellular debris, considerable development work was required in order to refine it.

3.3 Overcoming Reaction-Diffusion Limitation: Surface Catalysed Hygiene

An alternative approach to the delivery of active biocide onto colonised surfaces, is where the active biocide is generated, *in-situ*, by surface catalysis. The robustness of surface catalysed hygiene was demonstrated by Wood *et al.*[47] where reactive oxygen species were generated from persulphates and peroxides by transition metal catalysis at the colonised surface. The significant feature of this strategy is that it not only effectively overcomes reaction-diffusion limitation imposed on the access of many biocides but it also creates a diffusion pump by which further treatment agent is forced to the biofilm substratum interface. Since the catalysts are not consumed during the generation of active agent then the addition of treatment agent to the exterior of the biofilm will replenish the biocidal action at the sub-stratum. Furthermore, with peroxides and persulphates as the treatment agents, the treatments have been shown to result in a significant weakening of biofilm attachment and to promote its removal from the surface.

One application of this strategy incorporates transition metal catalysts, such as cobalt phthalocyanine and copper phthalocyanine, into the material that comprises the target surface. Catalysts, such as these break down peroxides and persulphates to liberate active oxygen species. Susceptibility of *Pseudomonas aeruginosa* biofilms towards potassium monopersulphate and hydrogen peroxide was enhanced significantly in all cases where such catalysts were incorporated in a trylon resin coupon. In all instances, catalysed killing of biofilm bacteria occurred at concentrations of treatment agent orders of magnitude less than those required for killing planktonic bacteria. Cobalt sulphonated phthalocyanine was more effective both as a catalyst for the decomposition of peroxide and as a hygiene enhancer than was copper sulphonated phthalocyanine. Significant improvements in delivered hygiene were apparent even with the relatively thick biofilms (100 μm).[48] Provided that catalytic activity is maintained and fresh treatment agent is available, such an approach ought to continually provide active biocide to the interactive surface between the biofilm and the substratum.

Suggested developments of this approach have been to utilise enzymes or enzyme combinations rather than inorganic catalysts. For example if enzyme combinations such as glucose oxidase and haloperoxidase were coated onto tooth surfaces or upon oral prostheses then hygiene might be delivered by the bodies own supply of treatment agent in the form of glucose and chloride. Significant enhancements of this form of oral hygiene might then be obtained by sucking a boiled sweet.

3.4 Biological Mimicry

Biological systems have evolved for the control of microbial fouling over many millions of years are often particularly effective. For example, the respiratory and gastrointestinal mucosa of mammals and various marine flora and fauna employ very effective colonisation control mechanisms.[49] Biological mimicry as a strategy for biofilm prevention relies on our ability to translate natural processes into artificial systems. Importantly, by their nature, natural biofilm-control mechanisms tend to be environmentally compatible and in this respect preferable to the profligate use of antimicrobial compounds. Although it could be argued that erodable surface coatings (above) are a form of biological mimicry in that sloughing of mucosal cells is an important protection mechanism in nature,[50] there are a number of emerging approaches which mimic other, more specific defence strategies.

Considerable interest has been expressed in the industrial use of stabilised hypothatous acids (water reacted with chlorine, bromine or iodine). This innovation imitates the stabilisation of oxidised bromide that occurs in natural systems.[51] These occur as mechanisms of control on the surface of some aquatic plants in the mammalian immune defences.[52] Certain marine algae produce hypobromous acid using bromoperoidases[53] which is not only an effective mechanism but exhibits good specific toxicity.

Another promising biomimetic strategy is the use of anti-signalling chemicals, from marine algae, to interfere with bacterial quorum sensing. Gram *et al.*[54] have demonstrated the use of such naturally occurring anti-signals (furanones) to interfere with the normal swarming motility of *Pseudomonas mirabilis* which is also regulated through homoserine lactone. This discovery of anti-quorum-sensing chemicals such as these[55,56] heralds the possibility of preventing adoption by bacteria of the biofilm phenotype and of thereby preventing resistance expression, regardless of its cause. Indeed, Davies *et al.*[31] have now demonstrated that signalling mutants of *P. aeruginosa*, deficient to varying extents in homoserine-lactone, produced biofilms that were devoid of extracellular polymeric materials. In contrast to wild-type biofilms these were sensitive to sodium dodecyl sulphate (SDS). This infers that homoserine lactones are involved not only in the adoption of attachment phenotypes but that they are also involved in regulation of biofilm exopolymers.

4 CONCLUSIONS

Microbial biofilms are responsible for considerable morbidity and mortality in medicine and, in conjunction with their commercial implications, are responsible for the world-wide loss of billions of pounds yearly. From a microbial physiologist's perspective, biofilms are a fascinating subject. Much of this fascination relates to their recalcitrance towards antimicrobial treatments which whilst effective against fast-growing, planktonic populations of individual species have failed spectacularly when deployed against biofilms. Considerable research effort and expense has gone into the development of new strategies for biofilm prevention, control and eradication. Biofilm drug resistance can be attributed to a variety of physiological properties, each of which have become targets for the development of novel treatment agents / strategies.

In this article, we have considered some of the more promising developments. Biocide-containing surfaces have demonstrated only limited success. The reason for this is that their effectiveness relates to the release of biocide. Whilst local concentrations of antimicrobial agent might protect the surface, leaching means that it is only short lived. Furthermore, microorganisms remote from the surface will be exposed to continuous, sub-lethal levels of biocide. Following leaching and depletion of biocide within the substratum, organisms deep within the biofilm may also receive sub-lethal doses. There is currently much concern that such exposures may lead not only to losses in the effectiveness of the incorporated agents but also to indirect effects upon the activity of third-party, therapeutic agents.[57,58] Whilst erodable, biocide-containing surfaces may to some extent overcome the problems associated with fouling of the substratum, they still require further development and will by their very nature suffer from depletion of agent and the indiscriminate exposure of organisms.

Perhaps the most promising future control methods are surface catalysed hygiene, where relatively innocuous treatment agents may be deployed, and biological mimicry, where toxicity is very specific and effects upon the environment are minimal. From an industrial perspective there is clearly no place in the new millennium for inefficient, indiscriminate methods that challenge the environment with toxic chemicals or potential inducers of drug and biocide resistance.

References

1. J. W. Costerton, G. G. Geesey and K-J Cheng, *Sci. Amer.*, 1978, **238**, 86.
2. J. W. Costerton, K. J. Cheng, G. G. Geesey, T. I. Ladd, J. C. Nickel and M. Dasgupta, *Ann. Rev. Microbiol.*, 1987, **41**, 435.
3. P. Gilbert and D. Allison, *Sci. Prog.*, 1992, **76**, 305.
4. J. M. Bernhard and S. S. Bowser, *Foraminifera Mar. Ecol-Prog. Seri.*, 1992, **83**, 263.
5. B. J. Paul, H. C. Duthie and W. D. Taylor, *J. N. Amer. Benthol. Soc.*, 1991, **10**, 31.
6. P. A. Holden, J. R. Hunt and M. K. Firestone, *Biotech. Bioeng.*, 1997, **56**, 656.
7. C. White, A. K. Sharman and G. M. Gadd, *Nat. Biotechnol.*, 1998, **16**, 572
8. S. Macfarlane, A. J. McBain and G. T. Macfarlane, *Adv. Dent. Res.*, **11**, 59.
9. J. A. Yu, M. Ji and P. L. Yue, *J. Chem. Technol. Biotechnol.*, 1999, **74**, 619.
10. R. F. Goncalves, V. L. DeAraujo and C. A. L. Chernicharo, *Water.Sci. Technol.*, 1998, **38**, 189.
11. M. Y. Lu, I. S. Maddox and J. D Brooks, *Process. Biochem.*, 1998, **33**, 117.
12. W. G. Characklis, in *Biofilms*, ed. W. G. Characklis and K. C. Marshall, Wiley, New York, 1990, p. 523.
13. B. J. Little, P. A. Wagner, W. G. Characklis and W. Lee, in *Biofilms*, ed. W. G. Characklis and K. C. Marshall, Wiley, New York, 1990, p. 635.
14. J. T. Holah, S. F. Bloomfield, A. J. Walker and H. Spenceley, in *Bacterial Biofilms and their Control in Medicine and Industry*, ed. J. T. Wimpenny, W. W. Nichols, D. Stickler and H. Lappin-Scott, Bioline, Cardiff, 1994, p. 163.
15. P. J. Eginton, J. Holah, D. G. Allison. P. S. Handley and P. Gilbert, *Lett. Appl. Microbiol.*, 1998, **27**, 101.
16. A. McBain, D. G. Allison and P, Gilbert, *Biotech, Bioeng, Rev.*, 2000, **17**, in press.
17. M. P. E. Slack and W. W. Nichols, *Lancet,* 1981, **11**, 502.
18. P. A. Suci, M. W. Mittelman, F. U. Yu and G. G. Geesey, *Antimicrob. Ag. Chemother.*, 1994, **38**, 2125.
19. P. S. Stewart, *Antimicrob. Ag. Chemother.*, **40**, 1996. 2517.
20. M. S. Favero, W. W. Bond, N. J. Peterson and E. H. Cook, in *Proc. Intern. Symp. Povidone.*, University of Kentucky, Lexington, 1983, p.158.
21. C. T. Huang, F. P. Yu, G. A. McFeters and P. S. Stewart, S. *Appl. Environ. Microbiol.*, 1995, **61**, 2252.
22. B. Giwercman, E. T. Jensen, N. Hoiby, A. Kharazmi and J. W. Costerton, *Antimicrob. Ag. Chemother.*, 1991, **35**, 1008.
23. M. Sondossi, H. W. Rossmore and J. W. Wireman, *Intern. Biodeter.*, 1985, **21**, 105.
24. P. Gilbert, in *Microbial Quality Assurance*, ed. M. R. W. Brown and P. Gilbert, CRC, New York, 1995, p. 61.
25. M. R. W. Brown, P. J. Collier and P. Gilbert, *Antimicrob. Ag. Chemother.*, 1990, **34**, 1623.
26. P. Gilbert, P. J. Collier and M. R. W. Brown, *Antimicrob. Ag.* Chemother., 1990, **34**, 1865.
27. P. Gilbert; D. G. Allison, D. J. Evans, P. S. Handley and M. R. W. Brown, *Appl. Environ. Microbiol.*, 1989, **55**, 1308.

28. P. Gilbert and D. G. Allison, in *Dental Plaque Revisited, Oral biofilms in Health and Disease*, ed. H. N. Newman and M. Wilson, Bioline, Cardiff, p. 125.

29. M. J. Ashby, J. E. Neale, S. J. Knott and I. A. Critchley, *J. Antimicrob. Chemother.*, 1994, **33**, 443.

30. J. R. Das, M. Bhakoo, M. V. Jones and P. Gilbert, *Appl. Microbiol.*, 1998, **84**, 852.

31. D. G. Davies, M. R. Parsek, J. P. Pearson, B. H. Iglewski, J. W. Costerton and E. P.Greenberg, *Science*, 1998, **280**, 295.

32. A. M George, and S. B. Levy, *J. Bacteriol.*, 1983, **155**, 531.

33. S. B. Levy, *Antimicrob. Ag. Chemother.*, 1992, **36**, 695.

34. D. Ma, D. N. Cook, M. Alberti, N. G. Pong, H. Nikaido and J. E. Hearst, *J. Bacteriol.*, 1993, **175**, 6299.

35. T. Maira-Litran, D. G. Allison and P. Gilbert, *J. Appl. Microbiol.*, In Press Sept. 1999.

36. P. Gilbert and D. Allison, in *Biofilms: The Good, the Bad and the Ugly*, ed. J. Wimpenny, P. Gilbert, J. Walker M. Brading and R. Bayston, Bioline, Cardiff, 1999, p. 263.

37. J. Rogers, J. B. Dowsett and C. W. Keevil, *J. Ind. Microbiol.*, 1995, **15**, 377.

38. D. K. Riley, D. C. Classen, L. E. Stevens and J. P. Burke, *Am. J. Med.*, 1995, **98**, 349.

39. D. J. Stickler and C. Winters, in *Bacterial Biofilms and their Control in Medicine and Industry*, ed. J. Wimpenny, W. Nichols, D. Stickler, H. Lappin-Scott, Bioline, Cardiff, 1994, p. 97.

40. S. K. Schmitt, C. Knapp, G. S. Hall, D. L. Longworth, J. T. McMahon and J. A Washington, *J. Clin. Microbiol.*, 1995, **34**, 508.

41. R. Bayston, in *The Life and Death of Biofilm*, ed. J. Wimpenny, P. Handley, P; Gilbert and H. Lappin-Scott, Bioline, Cardiff, 1995, p. 149.

42. C. Stanton, and R. Bayston, in *Biofilms: The good, the bad and the ugly*, ed. J. Wimpenny, P. Gilbert, J. Walker M. Brading and R. Bayston, Bioline, Cardiff, 1999, p.65.

43. D. J. Stickler, T. Williams, C. Jarman, N. Howe and C. Winters, in *The Life and Death of Biofilm*, ed. J. Wimpenny, P. Handley, P; Gilbert and H. Lappin-Scott, Bioline, Cardiff, 1995, p. 119.

44. K. E. Cooksey and B. Wigglesworth-Cooksey, in *Biofilms: Science and Technology*, ed. L. F. Melo, T. F. Bott, M. Fletcher and B. Capdeville, Kluwer Academic Publishers, Dordrech, 1992, p. 529.

45. C. Holmstrom and S. Kjelleberg, The effect of external biological factors on settlement of marine invertebrate and new antifouling technology. *Biofouling*, 1994, **8**, 147.

46. S. Suzanger, D. Allison, I. Eastwood and P. Gilbert, in *Biofilms: The Good, the Bad and the Ugly*, ed. J. Wimpenny, P. Gilbert, J. Walker M. Brading, M and R. Bayston, Bioline, Cardiff, 1999, p. 53.

47. P. Wood, M. Jones, M. Bhakoo and P. Gilbert, *Appl. Environ. Microbiol.*, 1996, **62**, 2598.

48. P. Wood, D. E. Caldwell, E. Evans, M. Jones, D. R. Korber, G. M. Wolfhaardt, M. Wilson and P. Gilbert, *J. Appl. Microbiol.*, 1998, **84**, 1092.

49. D. J. Stickler, in *Biofilms: Community Interactions and Control*, ed. J. Wimpenny, P. Handley, P. Gilbert, H. Lappin-Scott and M. Jones, Bioline, Cardiff, 1997, p. 215.

50. A. M. Shamsuddin, in *Gastrointestinal and Oesophageal Pathology*, ed. R. Whitehead, Churchill Livingstone, Edinburgh, 1989, p. 41.

51. W. F. McCoy, *Mater. Perform.*, 1998, **37**, p.45-48
52. E. L. Thomas, *J. Biol. Chem.*, 1995, **270**, 2906.
53. R. Wever, M. G. M. Tromp, B. E. Krenn, A. Marjani and M. Vantol, *Environ. Sci. Technol.*, 1991, **25**, 446.
54. L. Gram, R. de Nys, R. Maximilien, S. Giskov, P. Steinberg and S. Kjellenerg, *Appl. Environ. Microbiol.*, 1996, **62**, 4284.
55. R. De Nys, P. D. Steinberg, P. Willemsen, S. A. Dwarjanyn, C. L.Gabelish, R. J. King, *Biofouling*, 1995, **8**, 259.
56. M. Givskov, R. de Nys, M, Manefield, L. Gram, R. Maximilien, L. Eberl, S. Molin, P. Steinberg and S. Kjelleberg, *J. Bact.*, 1996, **178**, 6618.
57. L. M. McMurry, M. Oethinger and S. B and Levy, *FEMS Microbiol. Lett.*, 1998, **166**, 305.
58. L. M. McMurry, M. Oethinger and S. B and Levy, *Nature*, 1998, **394**, 531.

A NEW ENVIRONMENTALLY SENSIBLE CHLORINE ALTERNATIVE

William F. McCoy
Nalco Chemical Company
Naperville, IL 60563
wmccoy@nalco.com

1 INTRODUCTION

Far more chlorine is used to control microbial fouling compared to any other chemical. An environmentally sensible chlorine alternative is needed because handling the gas is hazardous, the liquid is not stable, combined residuals are not effective, free residuals do not control biofilms, and disinfection by-products are toxic. These problems have now been largely overcome for industrial water treatment with the introduction of a novel (the first) and unique (the only) stabilized liquid bromine antimicrobial. Although other stabilized halogen antimicrobials are well known in water treatment, none have substantially overcome the fundamental application problems of chlorine and bromine. This new product is the first biomimetic industrial antimicrobial, having been designed to imitate the N-bromoaminoethanesulfonate antimicrobial produced in the mammalian immune response.

The new antimicrobial is an order of magnitude less toxic, several orders of magnitude less volatile, easier to handle, more compatible with other water treatment chemicals, more effective against biofilms, and it generates less than half the disinfection by-products compared to chlorine or other alternatives. One hundred fifty billion gallons of industrial water have by now been successfully treated globally. Use of this new antimicrobial has substantially reduced environmental and human health risks from industrial water treatment by replacing nearly thirty million pounds of chlorine. The new product is proven to comparatively perform better, more safely, and it is substantially easier to apply than chlorine.

Improving the way industry manages microbial fouling does not depend entirely upon innovative new products. Since 1996, more than two thousand water treatment field engineers from every industrialized country and from many developing countries on every continent (except Antarctica) have been uniquely trained in microbial fouling control based upon sound principles in environmental microbiology. The foundation for success in this original approach to field practice is the industrial imitation of microbial fouling control strategies observed in nature. Field practice is arranged into three categories of key principles in natural microbial fouling control: Recognition, Remedy, and Regulation.

Recognition refers to the discovery of microbial fouling problems and then, determination of the value to be derived from solving root-causes of these problems. This requires use of modern diagnostic techniques in environmental microbiology and application of the science. Remedy refers to actions taken to minimize or eliminate root-causes of microbial fouling problems. This requires focus of the goal onto managing the overall microbial fouling process. Attempts to simply treat symptoms of the problem, such as has often been done in the past, are shown to be economically unacceptable. Regulation refers to proactive variance of the applied remedy so that the stresses applied will effectively control, for example, the complex surface fouling communities of microorganisms so often found to be at the root-cause of important industrial water treatment problems. Complex communities of microorganisms adapt frankly and quickly to the environmental stresses caused by applied industrial antimicrobials. Environmentally sensible strategies to manage the root cause of industrial microbial fouling problems must include proactive variance of the stresses applied. This practical approach to industrial water treatment has been readily accepted because of its compelling feasibility, its basis in environmentally-sound principle, and because the marketplace value of success is tangible.

2 IMITATING NATURAL MICROBIAL CONTROL

Natural microbial fouling control strategies are environmentally sensible because they have been optimized by natural selection. A sensible innovation strategy then, is to observe natural control, try to understand it, attempt an imitation, and explain the copy[1]. The new chlorine alternative and its industrial water treatment applications were accordingly developed, as follows.

Red and brown seaweed and also freshwater environments make hypobromous acid (HOBr) by selectively oxidizing bromide with hydrogen peroxide and an enzyme called bromoperoxidase (Figure 1). This oxidation was first observed in 1926; since then, fascinating scientific literature on the subject has developed. In one such study published in 1991[2], the brown seaweed *Ascophyllum nodosum* was shown to produce substantial quantities of hypobromous acid directly upon its surface. This aquatic plant may annually produce as much as 1,800 kg (3,960 lbs) of HOBr in the top few centimeters of the 30 km (18 mile) North Atlantic stretch of seawater studied. Many other macroalgae similarly produce hypobromous acid including the brown seaweeds (*Laminaria digitata, Fucus vesiculosis, Pelvetia canaliculata*) and also the red seaweeds (*Antithamnionella sarniensis, Antithamnion plumula, Chondrus crispuys, Placamium hamatum*). An example of *L. digitata* is shown in Figure 2.

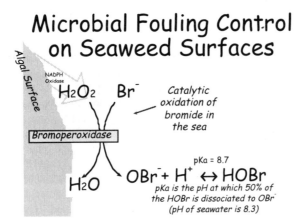

Figure 1 *The microbial fouling process on surfaces of certain macroalgae in aquatic environments is controlled by the selective oxidation of bromide with hydrogen peroxide and bromoperoxidase. Although chloride is many orders of magnitude more abundant in the sea, bromide is oxidized to hypobromous acid in situ.*

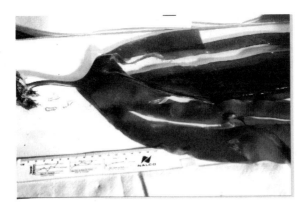

Figure 2 *The brown seaweed* <u>Laminaria digitata</u> *obtained from the North Atlantic off the coast of Newfoundland, Canada.*

Hypobromous acid so produced from the naturally occurring bromide in the ocean is rapidly consumed upon demand and is nontoxic to the plant. This natural process helps explain why seaweeds in the ocean are not overcome by adherent microbial biomass and macrofoulants that, for example, rapidly foul the surfaces of ships and piers in the sea. Although chloride is far more abundant than bromide, certain seaweeds preferentially make hypobromous acid *in situ*. The brominating activity of *Laminaria digitata* is shown by demonstration (Figure 3) in a replication of the work by Wever, et al.[2].

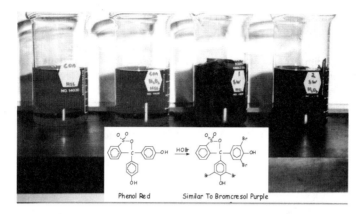

Figure 3 *Bromination of phenol red by* <u>Laminaria digitata</u> *in seawater (see Wever[2]). Beakers from left to right: 1) seawater plus phenol red, 2) same as 1 plus H_2O_2, 3) same as 1 plus* <u>Laminaria</u>, *4) same as 2 plus* <u>Laminaria</u>. *Phenol red is converted by the bromination.*

Superior antimicrobial activity in alkaline pH (seawater is always above pH 8), in the presence of nitrogenous organic matter, and due to lower volatility has been documented for bromine antimicrobials[3,4]. The pKa acid dissociation constants for HOCl and HOBr are 7.4 and 8.7, respectively; the dissociated acids are less effective antimicrobials[4,5]. Undissociated hypohalous acids are more effective because they are far better halogenating agents compared to the dissociated anion (hypohalite). Table 1 shows the effect of acid dissociation on antimicrobial performance in well-controlled laboratory experiments.

Table 1. *The effect of acid dissociation on antimicrobial performance against* <u>Pseudomonas aeruginosa</u> *at equivalent oxidant concentrations and contact times*

	Chlorine			Bromine	
pH	% HOCl	Log (bacteria killed/ml)	pH	% HOBr	Log (bacteria killed/ml)
7.0	79	5.8	7.0	99	5.9
7.5	50	4.8	7.5	98	5.9
8.0	30	3.5	8.0	90	5.9
8.5	12	<1	8.5	65	5.8
9.0	4	<1	9.0	40	4.0

These data show that bromine works better than chlorine in high pH waters such as the ocean. Similarly, most industrial water is quite alkaline and therefore, a practical form of bromine is also preferred. The technical attributes of bromine antimicrobials are of value in water treatment and are apparently also worth the cost to many aquatic plants. Further observations of natural microbial fouling control systems reveal that animals also preferentially manufacture, *in situ*, certain bromine-based antimicrobials.

The mammalian immune system protects the body from infection by many complex strategies. The most vigorous defense is performed by white blood cells known as granulocytes. These cells consume oxygen in response to microbial infections. This oxidative process, called the respiratory burst, has recently been proven to produce stabilized hypochlorite antimicrobials (predominantly by neutrophils) and stabilized hypobromite antimicrobials (predominantly by eosinophils)[6,7,8]. *In situ* oxidation of chloride and/or bromide by mammalian white blood cells occurs from the intracellular generation of hydrogen peroxide and selective halide oxidation via haloperoxidase catalysis. Intracellularly generated hypohalous acids then immediately react with stabilizers. These stabilized halogen antimicrobials are more effective and are less cytotoxic to animal tissues. Cationic membrane surfactants (bioactive peptides) and lytic enzymes that degrade bacterial cell walls and improve the efficacy of stabilized halogens are also simultaneously released from granules in a process known as degranulation. Figure 4 schematically shows the process as it occurs in eosinophils.

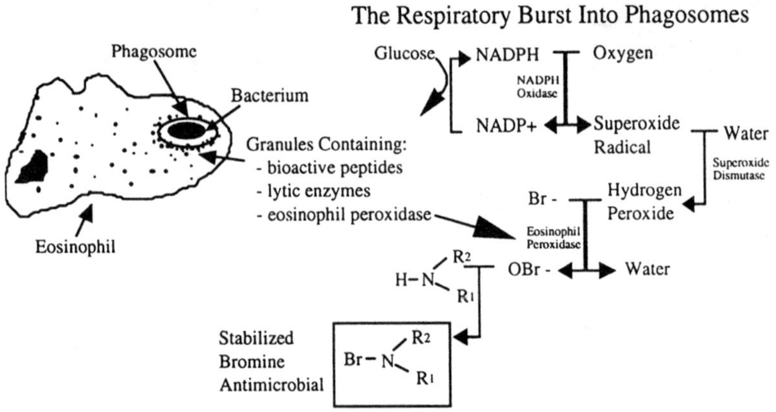

Figure 4 *Stabilized bromine antimicrobials are produced by eosinophils, a type of mammalian white blood cell. Bacteria are captured by phagocytosis and contained intracellularly within vesicles called phagosomes. Granules release cationic surfactants, lytic enzymes, and eosinophil peroxidase into the phagosome in a process known as degranulation. Eosinophil peroxidase, an enzyme that is structurally similar to the bromoperoxidases found in seaweed (Figure 1), selectively catalyzes oxidation of bromide to hypobromite by reducing hydrogen peroxide to water. The hypobromite immediately reacts with nitrogenous stabilizers such as aminoethanesulfonic acid (taurine) to form more effective and less toxic antimicrobial agents.*

Natural strategies to control microbial fouling often combine hypohalous acid disinfectants with nonoxidizing antimicrobials such as halogenated electrophilic organics, antibiotics, cationic peptide surfactants, and lytic enzymes. The industrial use of oxidizing biocides together with nonoxidizing biocides and biodispersants imitates this strategy and is often necessary because of the complexity and adaptability of the microbial fouling process.

2.1 The New Industrial Antimicrobial

A new industrial chlorine alternative was purposefully designed to imitate the stabilized bromine antimicrobials produced naturally in the human immune system. It is the first biomimetic industrial biocide. It is chemically analogous to the antimicrobial product of the oxidative respiratory burst in eosinophils, a type of mammalian white blood cell. In eosinophils, HOBr from the enzymatically-catalyzed oxidation of bromide with H_2O_2 immediately reacts with 2-aminoethanesulfonic acid (taurine). The product of this stabilization reaction is a potent antimicrobial, N-bromo-aminoethanesulfonic acid and it is the design model for the new industrial antimicrobial.

Amidosulfonic acid (sulfamic acid) is used to stabilize bromine in the patented proprietary process to manufacture and to use the new product[10]. Sulfamic acid is the reaction product of ammonia and chlorosulfonic acid, or it can be obtained by heating urea and sulfuric acid. Proprietary use of sulfamic acid in industrial water treatment as a means to protect organic water treatment chemicals from the oxidative effect of halogens has been previously disclosed[11]. Sulfamic acid is generally recognized as safe (GRAS) in CFR21 186.1093 for use as an indirect food ingredient.

STABREX® Stabilized Liquid Bromine[9] is far more stable than liquid chlorine bleach. For example, several tons of the new product were shipped to India and stored for one year above 90 °F. The product remained within specification (less than 10% degraded) for the entire year, after which it was successfully used to control fouling in an industrial water system. Chlorine would have completely degraded in this time under these conditions. Chemical wastage was eliminated. Accident risk in transporting oxidant was reduced because less volume was necessary. Table 2 shows the stability of the new product compared to industrial strength chlorine bleach in well-controlled laboratory tests.

Table 2 *Storage stability of chlorine and stabilized liquid bromine*

% Oxidant Remaining After Storage @ 77°F			% Oxidant Remaining After Storage @ 131°F		
Storage Days	Industrial Bleach	STABREX	Storage Days	Industrial Bleach	STABREX
0	100.0	100.0	0	100.0	100.0
12	90.2	98.3	1	76.8	96.6
32	83.9	98.3	3	40.2	96.6
135	59.2	97.4	7	26.8	86.2
246	43.4	96.4	12	16.1	75.9
600	23.5	95.2	20	8.9	65.5
800	18.7	94.0	33	5.4	53.4

Unstabilized hypobromite solutions are even more unstable than hypochlorite. Hypobromite disproportionates to bromate (BrO_3), a toxic and potentially carcinogenic compound, in alkaline conditions. The stabilizer in STABREX inhibits that process as shown in Table3.

Table 3 *Stability to bromate formation*

Oxidant (Equivalent Concentrations)	% Bromate After 60 Days Storage
Sodium Hypobromite	2.7
Stabilized Liquid Bromine	0.004

Volatility is the key technical problem with many chlorine alternatives such as ozone and chlorine dioxide because most industrial systems are open to the atmosphere and the treated water is vigorously recirculated. The volatility of STABREX is lower than any other oxidizer used for industrial water treatment as shown in Table 4.

Table 4 *Relative volatility of oxidizing chemicals*

Oxidizing Chemical	Air-Water Partition Coefficient (Atm @ 20°C)	Relative Volatility (Normalized)
Ozone	5000	125,000
Chlorine (elemental)	585	15,000
Bromine (elemental)	59	1475
Chlorine Dioxide	54	1350
Hypochlorous Acid	0.08	2
Hypobromous Acid	0.04	1
STABREX	<< 0.04[a]	<< 1

[a]Volatility was below the limit of measurement in the method used

Chlorine is highly reactive with scale and corrosion inhibitors and reactions are counter-productive. STABREX is less aggressive to many inhibitors such as those indicated in Table 5.

Table 5 *Compatibility of oxidizing antimicrobials with water treatment compounds*

Oxidizing Chemicals[a]	AMP[b] % Reverted ± std dev	TT[c] % Degraded ± std dev
None	0.9 ± 1.3	0 (normalized)
Chlorine	39.0 ± 2.8	18.1 ± 0.1
Bromine	20.7 ± 1.3	96.6 ± 1.3
STABREX	4.1 ± 1.9	2.2 ± 4.6

[a] 5 ppm total residual oxidant (measured) for 1 hr. contact
[b] AMP is aminotrimethylene phosphonic acid and is useful as a scale inhibitor
[c] TT is tolyltriazole and is useful as a yellow metal corrosion inhibitor

Lower chemical reactivity with non-target molecules is useful for another performance-related reason. Microorganisms prefer the protection and luxuriant environment in biofilms (the adherent microbial communities that cause detrimental surface-fouling effects in water cooling systems). Most (>99%) of the viable microorganisms in industrial systems are found in biofilms, not floating around freely in the bulk recirculating water. Compared to unstabilized chlorine or bromine, STABREX more effectively removes and disinfects biofilms as shown in Table 6.

Table 6 *Performance of equivalent bromine antimicrobial treatments against biofilms in well-controlled laboratory experiments to simulate industrial applications*

Oxidizing Chemical	% Removal of Biomass (± std dev)	% Decrease in Fluid Frictional Resistance (± std dev)	Biofilm Disinfection (Log Reduction in Viable Bacteria/cm²)
Unstabilized Bromine	2 ± 2	3 ± 3	3.4 ± 0.2
STABREX	45 ± 2	47 ± 18	5.2 ± 0.2

The new stabilized bromine antimicrobial is an excellent antimicrobial having been proven superior in field and laboratory experiments compared to chlorine, stabilized chlorine, and equal to or better than solid hypobromite antimicrobials. The product is effective for the control of microbial biofilms and highly diverse microbial communities, including those that harbor *Legionella*[5].

STABREX is safer to use because it is less toxic to aquatic wildlife, as shown in Table 7, and because less chemical is required to control microbial fouling.

Table 7 *Toxicity comparison of unstabilized and stabilized bromine in standardized aquatic toxicity tests*

Aquatic Toxicity Test (LC50) [a]	Unstabilized Bromine (ppm as Br_2)	STABREX (ppm as Br_2)	Toxicity Reduction
Rainbow Trout (96h LC50)	0.23	0.60	2.6x
Daphnia magna (48h LC50)	0.04	0.58	14.5x
Sheepshead Minnow (96h LC50)	0.19	2.30	12.1x
Mysid Shrimp (96h LC50)	0.17	3.65	21.5x

[a] LC50 is the lethal concentration for 50% of the test organism after 96h or 48h contact

Disinfection by-products (e.g., adsorbable organic halides such as trihalomethanes) are more than 50% decreased compared to equivalent chlorine treatments in standardized AOX test with STABREX[5]. In practice, disinfection by-products are decreased even further in STABREX applications because less oxidant is required to control the microbial fouling process compared to bromine or chlorine applications.

STABREX is easier and simpler to use compared to any other oxidant available for industrial water treatment. The product is pumped directly from returnable transporters (PortaFeed® Systems)[17] with standard chemical feed equipment. Previously, the only practical ways to apply bromine were to oxidize bromide solutions on-site with chlorine in dual liquid feed systems, or with one of the solid organically-stabilized bromine products applied from sidestream erosion feeders. The former is cumbersome and complex, and the latter is prone to dusting and difficult to control. Other oxidants require complex handling and feed of toxic volatile gases, unstable liquids, multiple-component products, or reactive solids. Simplicity in use results in reduced risk to workers and to the environment.

2.2 Industrial Benefits of the New Antimicrobial

The benefits of replacing chlorine with STABREX are in reducing environmental toxicity (because it is less toxic to aquatic wildlife), in reducing accident risk (because it is less hazardous and easier to handle), and in reducing chemical waste (because it works better, is more stable in transport/storage, is less volatile and less reactive). Environmental toxicity and accident risk have been substantially reduced in more than 2,500 industrial water systems worldwide.

Significant industrial demand for a more practical stabilized liquid bromine product has been known for several years. Until the invention and commercial development of STABREX, there was no practical means to overcome the inherent instability, volatility, and handling hazards of liquid bromine. The new technology solves several long-standing technical problems that bromine manufacturers tried, unsuccessfully, to overcome for 15 years. The practical use of STABREX stabilized liquid bromine and its successful application in microbial fouling control is new to the world of industrial water treatment.

3   IMITATING NATURAL PROGRAMS

Natural microbial fouling control strategies consist of much more than just effective biocides. In fact, natural strategies look like what we would call programs. They are comprised of systems to recognize the problem, remedy it, and then regulate the remedy. Practicing these three (recognition, remedy, and regulation) is necessary to effectively manage industrial microbial fouling.

3.1 Recognition

Natural strategies to control the microbial fouling problem always depend upon diagnosis and monitoring. Natural recognition systems are far more sophisticated than the current state-of-the-art in our industry. For example, eosinophils capture and destroy bacteria that have been marked with compounds such as specific antibodies from the immune response. These substances, called opsonins, stimulate eosinophils to capture the invading microbes, activate the release of granule contents (degranulation) and then, trigger the respiratory burst.

Effective industrial programs begin with establishing criteria for success. These criteria are the result of diagnostics (e.g., surveying the total system to find root causes of microbial fouling problems) and monitoring (e.g., trended quantitative measurements of microbial fouling indicators related as directly as possible to the fouling process). Frankly, many spectacular program failures could have been avoided if the root cause of microbial fouling problems had been recognized earlier.

Practical methods are now available for diagnostics and monitoring but more research, innovation, and commercial development presents a very great potential to improve the performance of industrial programs. Several new innovations have been recently introduced such as an effective optical fouling-deposit monitor[12].

3.2 Remedy

Some of the preferred tools used in natural microbial control programs are given in Table 8. The goal in nature is always to maintain control of a system, to avoid letting it foul until cleanup is the last resort. In nature, maintenance of a "clean" system is the only real hope for survival. Clean rarely, if ever, means sterile. The natural systems discussed in this paper, for example, are always comprised of vast microbial flora in close proximity to or actually a part of the protected portion of the system in which microbial fouling problems are actively managed by the animal or plant.

Similarly in water treatment, management of the overall microbial fouling process is the only strategy that works. Sterilization is seldom the goal in water treatment; minimizing the microbial diversity and maintaining control is the key. Management of the overall microbial community is certainly the best means to reduce the risk of legionellosis, for example[5]. Microbial diversity is an important indicator of the extent to which control has been achieved[13,14].

Table 8 *Natural microbial control remedies, a few examples, and a few industrial copies.*

Natural Microbial Control Remedy	A Few Examples	Industrial Copy
Halogenating Agents	*In situ* halide oxidation by mammalian white blood cells, fungi, and aquatic plants	Stabilized Hypohalous Acids
Antimicrobials and Antibiotics	Specific metabolic inhibitors, electrophilic organics, cell signaling inhibitors	Nonoxidizing Biocides
Surfactants, Detergents, and Enzymes	Bioactive peptides, lytic enzymes, anionic surfactants	Biodispersants and Biodetergents

Diverse microbial communities adapt to stresses. The common control motif in nature is to vary the applied stresses on microbial communities using combinations of antimicrobials. The consequence of failing to do so is selection of an increasingly tolerant fouling community. This undesirable selection process will usually proceed until drastic cleanup remedies become necessary to save the system. To the authors' knowledge, natural microbial control programs never depend upon the exclusive unvarying use of a single antimicrobial.

3.3 Regulation

It is interesting to observe regulation of the remedy in natural microbial fouling control programs. Examples from nature are remarkably elegant. For example, feedback inhibition of haloperoxidase synthesis in eosinophils is caused by stabilized bromine antimicrobials, thereby limiting the concentration of halogenating agent that can be produced. Hypobromous acid actually inhibits the enzymatic oxidation of halide by bromoperoxidase in seaweed. Thusly, regulation of the remedy protects animal and plant from excessive concentrations of biocide.

State-of-the-art, performance-based regulation of industrial biocides is a great opportunity for innovative improvement. Recent innovations include on-line performance-based monitoring, dosage control, and on-site biocide active analysis[12,15,16]. This new technology promises to improve performance efficiency and reduce the risk associated with managing the microbial fouling process.

4 CONCLUSION

The clear lesson from nature is that effective management of microbial fouling in complex dynamic systems requires a consistently administered program comprised of diagnostics, monitors, environmentally sensible remedies, and careful regulation of those remedies.

A sensible approach to innovation in microbial fouling control technology can be simply stated: Observe nature. Try to understand it. Try to imitate it. Explain the copy. There is much more to learn about natural microbial fouling control. Surely, there are many important clues still to be discovered.

On innovating to imitate nature: The wonders of nature inspire imitations that can yield environmentally sensible solutions to industrial problems.

5   ACKNOWLEDGMENTS

The author thanks Scott Borchardt for the demonstrations with *Laminaria digitata*. Thoughtful and stimulating discussions with Tony Dallmier, Shunong Yang, Eric Allain, and Michael Enzien are greatly appreciated. It has been a pleasure to work with these individuals and also Eric Myers, Rob Kelly, Philip Yu and Steve Bradley in the commercial development of this technology.

References

1.   McCoy, W.F. *Materials Performance*, 1998. **37**(4):45-48
2.   Wever, R., et al. *Environ. Sci. Technol.*, 1991, **25**:446-449.
3.   Blatchley, et al. *Water Research*, 1992. **26**:99-102
4.   McCoy, W.F., et al., *Cooling Tower Institute*, 1990, Paper No. TP 90-09.
5.   Dallmier, A.W, et al. *Natl. Assoc. Corrosion Eng*, 1997, Paper No. 398.
6.   Weiss, S., et al. *Science*. 1986. **234**:200-203.
7.   Mayeno, A.B., PhD Thesis. University of California, Los Angeles, 1989
8.   Thomas, E.L., et al., *J. of Biol. Chemistry*, 1995, **270**(7):2906-2913.
9.   STABREX® Stabilized Liquid Bromine is a trademark of Nalco Chemical Company
10. U.S. Patents 5,683,654, 5,795,487, and 5,922,745 assigned to Nalco Chemical Co.
11. U.S. Patents 4,642,194, 4,711,724, 4,759,852, 4,929,424, 5,589,106, 4,992,209, and 4,883,600 assigned to Nalco Chemical Company
12. Wetegrove, R.L., et al. *Cooling Tower Institute*, 1996, Paper No. TP96-09.
13. Cloete, T.E., et al. 1989. *Water SA*. 1989, **15**(1):37-41.
14. Brözel, V.S., et al. 1992. *Water SA*. 1992, **18**(2):87-92.
15. Borchardt, S.A., et al. *Natl. Assoc. Corrosion Eng.*, 1997, Paper No. 466.
16. McCoy, W.F., et al. *Cooling Tower Institute*, 1995, Paper No. TP95-16.
17. PortaFeed® is a registered trademark of Nalco Chemical Company, Naperville, IL

Paints and Coatings

PRESERVATION OF PAINTS IN THE WET-STATE

John Gillatt

Thor Specialities (UK) Limited, Earl Road, Cheadle Hulme, Cheshire SK8 6QP, UK

SUMMARY

Because of the raw materials that they contain and the conditions under which they are manufactured and stored, water based paints are highly susceptible to microbiological infection and spoilage in the wet-state.

The effects of microbial infection are viscosity loss, pH change, gassing, malodour and visible surface growth. In addition, discoloration and alteration in the rheology of the formulation can result in the product being unacceptable to the customer.

However, the use of ever increasing concentrations of more highly active biocides is not the complete answer to these problems.

Instead, the producer, with the help of the biocide manufacturer, must critically look at the whole production operation from a microbiological point of view.

Improvements in plant hygiene, including examination and possible modification of plant design and more frequent and thorough cleaning, including the use of a biocidal wash, will help to alleviate problems. Changes in production practices, such as avoiding storage of stock thickener solutions and careful monitoring of raw materials will bring about further improvements.

These activities, allied with the use of effective broad-spectrum biocides, will at least decrease incidences of microbiological contamination and at best eliminate such infections altogether.

In this paper the author will attempt to answer the questions:-

- What are micro-organisms?
- What are they called?
- Where do they come from?
- Why do they grow in paint?
- What effects can they produce?
- How can they be prevented?

Information will be given on a novel isothiazolinone combination biocide offering many advantages over traditional chloromethyl/methyl- and benz-isothiazolinone based preservatives.

1 WHAT ARE MICRO-ORGANISMS?

Biocides have been called the "necessary evil" but in reality they are as valuable to the manufacturer of susceptible products as the insurance on the plant or buildings. Like an insurance policy they are designed to give protection when things go wrong.

The risk that biocides insure against is that caused by infections of micro-organisms, the so-called "invisible enemy", and to appreciate the risks involved it's first necessary to learn something about this group of living creatures.

Micro-organisms (Figure 1), bacteria, moulds and yeasts, collectively known as fungi, are minute living entities, too small to be seen with the unaided eye.

Figure 1:   *Micro-organisms*

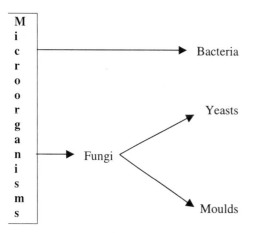

Bacteria and fungi are sufficiently different in their form, function, metabolism and growth requirements to pose quite different risks to formulated products.

1.1 Bacteria

Bacteria (Figure 2) are the main enemy of the water based paint manufacturer.

Figure 2:   *Bacteria*

- Single celled, variously shaped
- Very small (0.5 - 2.5µm, 10^{-12} grams)
- Require:-
    water
    food
    warmth
    correct pH
- Reproduce rapidly
    as often as every 20 mins

Because they grow so rapidly under ideal conditions (Figure 3) a low level of contamination of only a few hundred per gram can reach problem proportions in just hours.

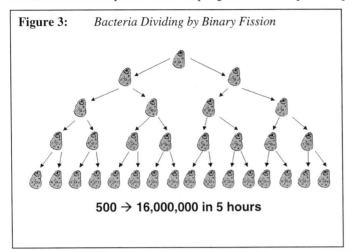

Figure 3:   *Bacteria Dividing by Binary Fission*

500 → 16,000,000 in 5 hours

Generally they prefer a temperature of between 20 and 40°C, a pH that is slightly alkaline and sources of carbon (food), trace elements and oxygen. Some bacteria do not need gaseous oxygen utilising it from inorganic sources such as sulphates and nitrates, liberating sulphur compounds, including H_2S and nitrites or even nitrogen.

1.2 Fungi

Fungi comprise two sub groups of organisms, the moulds and the yeasts.

Moulds (Figure 4) have complex filamentous structures and fruiting bodies often with highly coloured sporing heads. Their requirements for growth are similar to bacteria although they do need gaseous oxygen for growth. They also tend to prefer cooler temperatures, some grow better in the dark and all have a preference for a slightly acidic pH.

Figure 4:   *Moulds*

- Individual cells are very small (10□m in diameter)
- Require:-
    water
    food
    warmth
    pH
    oxygen
- Reproduce by spore production

1.3 Yeasts

The second group of fungi is the yeasts (Figure 5). These are similar in some respects to the bacteria, being small unicellular organisms that divide rapidly under ideal conditions (once every 80 minutes). However, within the micro-organisms they are classified as fungi because their metabolism and other properties are the same as the moulds.

They do differ in one respect from their filamentous siblings in that they are able to metabolise certain carbohydrates (sugars) by a process known as fermentation. During this, of course, they produce alcohol and gas (carbon dioxide) and for that reason they are used in brewing, bread making and other useful processes.

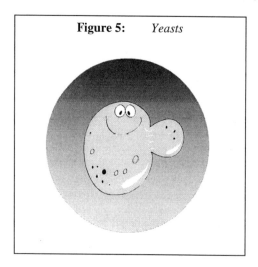

Figure 5:   *Yeasts*

2   MAIN CONTAMINATING ORGANISMS

Millions of species of micro-organisms have been discovered since the early pioneers of microbiology started to give them names and many thousands of different bacteria, moulds and yeasts have been found in a wide range of industrial water based products.

Of these, probably several hundred have been shown to actually cause contamination problems. Although the most important thing to be able to do is to cure and prevent microbial growth in products it is sometimes useful to know which organisms are growing in order that their origin can be determined.

For instance, if a paint is contaminated with six different bacteria and a defoamer used in the product contains the same organisms then it's probable that the defoamer is the cause of the problem.

A literature search will reveal many studies giving lists of contaminating organisms.

For example, a collation of results of five researchers in the 1970s and 1980s (Table 1) shows *Bacillus, Pseudomonas, Proteus, Enterobacter* and *Escherichia* to be the most commonly contaminating aerobic bacteria. In addition Opperman and Goll (1984) also investigated the incidence of anaerobic micro-organisms finding *Bacteroides, Clostridium, Desulphovibrio* and *Bifidobacterium* species.

Table 1:   *Bacteria from Emulsion Paints*

Species isolated ( )*		Huddart (1983)	Jakubowski et. al. (1983)	Miller (1973)	Opperman and Goll (1984)	Woods (1982)
Aerobacter	(2)			X		X
Aerococcus	(1)			X		
Achromobacter	(1)		X	X		
Alcaligenes	(1)	X		X		
Actinomycetes	(1)	X		X		
Bacillus	(8)			X		X
Citrobacter	(1)		X		X	
Enterobacter	(3)	X			X	
Escherichia	(2)	X		X	X	
Flavobacterium	(1)	X		X		
Hafnia	(1)				X	
Klebsiella	(1)				X	
Micrococcus	(2)	X		X		
Proteus	(3)	X		X	X	
Pseudomonas	(4)	X	X		X	X
Serratia	(2)					X
*( ) Number of different species reported						

Other workers including Jakubowski *et. al.* (1983) have found many different yeast and mould contaminants comprising most of the well known biodeteriogenic species such as *Aspergillus, Fusarium, Geotrichurn, Penicillium, Scopulariopsis, Saccharomyces* and *Torula*. In addition, the author has also isolated *Rhodotorula* and *Sporobolornyces* on a number of occasions.

3   SOURCES OF INFECTION

There are a number of main sources of microbial contamination (Table 2) of paints and other products that are affected by microbial contamination in the wet-state.

Table 2:   *Sources of Wet-State Microbiological Contamination*

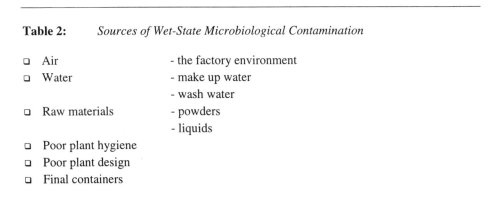

- ❑ Air   - the factory environment
- ❑ Water   - make up water
                                               - wash water
- ❑ Raw materials   - powders
                                               - liquids
- ❑ Poor plant hygiene
- ❑ Poor plant design
- ❑ Final containers

3.1 The Air

The air that we breathe is full of microbial cells and spores of bacteria and fungi. Because they are extremely light they are readily are carried by wind currents. In hot weather soil, a rich source of all types of microbes, turns to dust and increases the airborne microbial population

3.2 Water

Water is one of the most widely used raw materials in the manufacture of aqueous paint formulations. But as well as being used in production it can have a number of other uses, e.g. cooling and cleaning. Because of the cost of "new" water and disposal of contaminated water most manufactures have introduced closed loop or recycling systems.

New water may come from a number of sources (Table 3) and the degree of inherent contamination will vary accordingly.

Table 3:   *Water and its Contaminants*

Water from	Contamination
❑ Mains supply	-/+
❑ Boreholes/Wells	++
❑ Rivers	+++
- = no contamination to	+++ = heavy contamination

Mains drinking water is not sterile. Water regulations state that it must contain less than one faecal coliform bacterium per 100 cm3 but concentrations of 10^3/cm3 of *Pseudomonads*, one of the main causative organism in industrial product spoilage, are quite common and can lead to infection.

Because of the cost of mains water, many companies have utilised on-site boreholes as their water source. Such water, percolating through the soil, will be much more contaminated with micro-organisms than the mains supply and will require treatment before use. River or canal water will contain even more micro-organisms and, although after sand or other filtration may appear clear, it still represents a significant contamination hazard to products in which it is used.

The other main uses for factory water are for cooling and cleaning. Water used for cleaning the plant will contain trace amounts of residual manufactured product from the vessel being treated. This water represents diluted product, still containing enough nutrients for microbial growth but not enough biocide to prevent organisms from developing. Water left in pipework or incorrectly stored flexible hoses can pose a similar potential hazard.

Often wash or clean down water is used as the basis for the next batch of product to be manufactured and, if contaminated, will infect the new material.

3.3 Raw Materials

Water is not the only raw material to represent a contamination hazard.

Many powdered raw materials, e.g. china clay, talc and calcium carbonate are natural products, coming from the soil and often contaminated with appreciable numbers of bacterial and fungal spores. Products heated to high temperatures during preparation, e.g. calcined kaolin clay used as a titanium dioxide extender, will generally be free from such contaminants.

Contamination is not restricted to natural products. Organic pigment powders are often precipitated from aqueous solutions and washed with water prior to drying. These have been found to contain high numbers of microbes and have been implicated in contamination instances.

Table 4:   *Potentially Contaminated Raw Materials*

❑ Powders   - Fillers
                  - Extenders
                  - Pigments
❑ Liquids   - Polymer emulsions
                  - Defoamers
                  - Pigment dispersions
                  - Dye solutions
                  - Dispersing aids
                  - Emulsifiers

Liquid raw materials such as polymer emulsions, defoamers, pigment dispersions, dye solutions, dispersing aids and emulsifiers are all products that can themselves become infected with micro-organisms if not produced from non-contaminated ingredients, under good manufacturing conditions and with an effective preservative.

Even a low addition concentration of a heavily contaminated raw material can introduce significant levels of micro-organisms into a product. This can then provide the focus for a much more serious problem.

3.4 Plant Hygiene

It is said that cleanliness is next to godliness and manufacturing plants are no exception to this maxim. Many of the contamination problems that occur in manufactured products have their origins in poor plant hygiene, resulting in uncertain product, customer complaints and unnecessarily high biocide addition levels.

The trend from heavy metal and phenolic based biocides, e.g. mercury and pentachlorophenol types, to more environmentally acceptable but less persistent organic types, requires more attention to plant hygiene (Figure 6, Briggs, 1980).

The so-called "splash zone", the head space in a mixing vessel above the bulk phase of product, may be irrigated with condensation water, washing less persistent biocides away and allowing microbial growth to occur in this area.

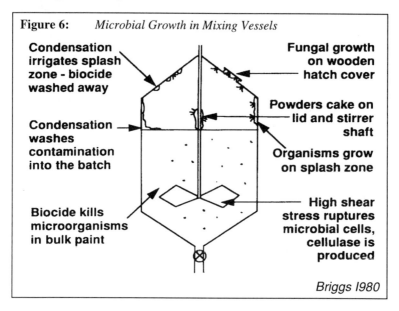

Figure 6:   *Microbial Growth in Mixing Vessels*

Condensation irrigates splash zone - biocide washed away

Condensation washes contamination into the batch

Biocide kills microorganisms in bulk paint

Fungal growth on wooden hatch cover

Powders cake on lid and stirrer shaft

Organisms grow on splash zone

High shear stress ruptures microbial cells, cellulase is produced

Briggs 1980

Wooden hatch covers may support and encourage growth, particularly of fungi, and powders, caked onto the stirrer shaft and lid, moistened with condensation and not biocidally protected, can also become colonised by micro-organisms.

Condensation will wash contaminating organisms onto the surface of the bulk phase where, especially if the vessel remains unstirred for some time, dilution of the surface layer, and hence the biocide in it, can allow profuse microbial growth to occur. When mixing recommences, such a high microbial loading will enter the bulk phase and may "overwhelm" a biocide.

Even if all of the organisms are killed off, their death and breakdown, either by the biocide or by stress rupture, especially if shear mixing is used, may release enzymes. These, unaffected by the biocide, will often be cellulolytic and can remain active for prolonged periods.

Most modern manufacturing plants are complex pieces of engineering. In fact it seems that many chemical engineers judge the success of a design not by its simplicity but by its complexity.

Long pipework runs, deadspots and sharp bends can all harbour contamination and are difficult to effectively clean.

Often old, unused pipework is cut off or blocked up leaving a small piece behind in which contaminated product can collect and infect new material passing through or stored. In cases like this the only solution is complete removal of the offending piece.

3.5 Final Containers

Good, microbiologically clean product can be spoiled at the final stage by filling into contaminated packaging. Metal cans, brought in from a cold warehouse to a warm factory can suffer from condensation and entrapment of airborne dust along with the microbes that contains.

Plastic containers often have residual mould release agents and can leach nutritious plasticisers into the product filled into them.

Quite often empty containers are stored in such a way that they are subject to the weather and to contamination that may result. A quick wipe with a dirty rag is all that is often used to clean them before filling! Containers really do need to be stored in a clean, warm environment in their original packaging to avoid this type of problem.

4   WHY MICRO-ORGANISMS GROW IN PAINT

4.1 Growth Requirements for Micro-organisms

The required growth conditions for bacteria and fungi are summarised in Table 5.

Table 5 *Growth Requirements for Micro-organisms*

Requirement	Bacteria	Moulds and Yeasts (Fungi)
❑ Light	✗	✗
❑ Ideal pH	slightly alkaline	slightly acidic
❑ Ideal temperature	25 - 40°C	20 - 35°C
❑ Nutrients	C, H, N sources	C, H, N sources
❑ Trace elements	✔	✔
❑ Oxygen	O_2 or inorganic, e.g. SO_4, NO_3	O_2
❑ Water	liquid or vapour	liquid or vapour

Most modern water-based paints, the raw materials they contain (Table 6) and the way in which they are manufactured and used provide ideal conditions for microbial growth.

The pH of most formulations is in the range 8 to 10, they usually have plenty of sources of carbon and trace elements and oxygen is dissolved in the water on which they are all based.

Most formulations reach, during manufacture or on storage, the ideal temperature for the growth of contaminants. If warm during production and then stored in insulated tanks they may remain above ambient temperature for some time.

As a result, low levels of microbiological infection can reach problematic concentrations overnight

Table 6: *Typical Emulsion Paint Formulation*

Thickener (cellulosic)
Defoamer
Coalescing agent
Dispersing agent
Extenders/fillers
Pigment
Polymer emulsion (binder)
Ammonia
pH 8 - 9
Water

5   THE EFFECTS OF MICROBIAL INFECTION

Small numbers of organisms will always be present in industrial plants and low levels of organisms, provided that they do not grow, do not generally cause problems.

However, because the growth requirements for bacteria and fungi are ideally found in water based paint formulations and because bacteria in particular reproduce so quickly, small numbers can rapidly reach problem proportions unless they are inhibited, e.g. by the use of a suitable biocide

The effects of microbial infection (Figure 7) are:

❑   Viscosity loss

❑   pH changes

❑   Gassing in the can

❑   Foul odours

❑   Discoloration

❑   Visible surface growth

❑   Enzyme Production

Figure 7:   *Effects of Microbial Infection*

Visible surface growth

Gassing

pH drift

Viscosity loss or increase

Malodour

Discoloration

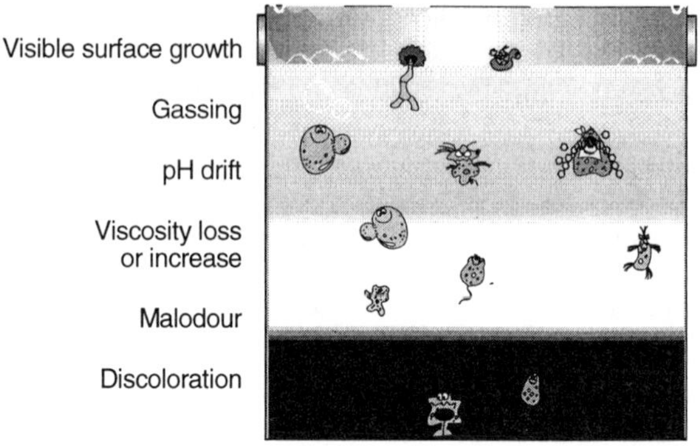

5.1 Viscosity Loss

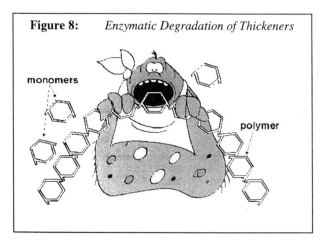

Figure 8:   *Enzymatic Degradation of Thickeners*

monomers

polymer

Viscosity loss (Figure 8) is brought about by enzymes, mainly cellulases, produced as organisms break down components such as cellulose ether thickeners.

In breaking down the long polymeric cellulose chains to shorter oligomers the organisms produce sugars that they can then metabolise further.

5.2 pH Changes and Gassing

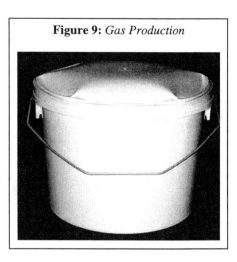

Figure 9: *Gas Production*

Breakdown of sugars results in the production of organic acids, which can reduce the pH of paints by more than one unit. Fermentative bacteria will produce gas from breakdown of the cellulosic thickener (Skinner, 1970) degrading the cellulose first to glucose which is then fermented to yield acid plus carbon dioxide (Stannier *et. al.*, 1971). In the presence of fillers. such as calcium carbonate, further gas can he produced by the action of the biogenic acids (Briggs, 1977). Such gas formation, not usually noticeable during production, can cause distortion and even splitting or "lid popping" of containers.

5.3 Malodours

Figure 10: *Bad Smells*

Some bacteria can give products a rancid smell; others can impart the "sweet" odour of dirty drains by the production of certain pyrazine derivatives. Other bacteria, known as sulphate reducers, for example *Desulphovibrio desulphuricans,* are able, under anaerobic conditions, to utilise oxygen from sulphates leading ultimately to the formation of hydrogen sulphide. Opperman and Goll (1984) in their study of contaminated emulsion paints concluded that more than a quarter were infected with these and other anaerobic organisms.

Such growth of sulphate reducing bacteria is responsible for the most commonly noticed malodour associated with emulsion paint spoilage. Hydrogen sulphide levels in paint have never been shown to have reached toxic concentrations, but even very small amounts render a product unsaleable.

5.4 Discoloration

Formation of insoluble sulphides from H_2S production by sulphate reducing bacteria can bring about blackening of products (Figure 11) and some bacteria such as *Serratia* and *Flavobacteria* species and yeasts, including *Rhodotorula* can give pink or yellow discolorations.   Other bacteria such as the *Pseudomonads* can produce fluorescent pigments.

Figure 11:   *Blackening of Paint*

5.5 Visible Surface Growth

Figure 12:   *Fungal Growth on Product Surface*

One other effect is caused mainly by fungi and that is visible and disfiguring growth on the surface of the product. (Figure 12).

This rarely occurs on its own, often happening once other organisms have reduced the normally alkaline pH of many products to a value at which these organisms can grow.

It's most commonly seen on the surface of high viscosity products such as artists' products and children's paints.

6   PREVENTION OF MICROBIAL INFECTION

An integrated approach to protection of emulsion paints must be taken if microbiological problems are to he avoided.

6.1 Raw Materials

Part of the company's quality control system should involve inspection of raw materials prior to use. Obviously contaminated products exhibiting one or more signs of microbial infection should not be used and any doubtful materials should be checked by the user, the supplier or a third party.

It may be possible to use less microbiologically susceptible raw materials. For example, hydroxypropyl methyl cellulose has been shown to be more resistant than hydroxyethyl and carboxymethyl cellulose (Briggs, 1980).

Water quality is very important and, as this is so variable, routine monitoring of microbiological contamination should be carried out. If necessary, water can he treated with the biocide used during manufacture, formaldehyde or a formaldehyde generator, or be passed through a UV sterilising system. Ion exchange columns and storage tanks are particularly problematic and these need to he kept physically and microbiologically clean at all times.

6.2 Production

Good plant hygiene is one of the key factors in avoiding problems.

Thorough clean downs should be carried out as often as the production schedule will allow, during which all accretions, build up of caked powders etc., are removed by scraping, power steam cleaning or alkali soak, back to the bare surface. Sterilisation can then he achieved by a biocidal rinse, this protected rinse water then being used for the next paint batch produced. Filling equipment, especially nozzles, must be similarly treated.

Biocide should also he added to any water overlay and to water or diluted product left in mixing vessels, storage tanks, pipelines and hoses.

Consideration can also be given to installing a humidifying system incorporating a biocide spray inside mixing vessels and storage tanks.

Pipcwork and transfer hoses should be as short as possible, free from sharp bends and deadspots and routinely cleaned and sterilised with a biocide solution. Hoses should be stored in such a way (Figure 13) as to allow them to drain and avoid becoming a source of infection.

Care must also he taken in the use and storage of the final containers to ensure that they are dry and free from contamination before filling and that, when filled, they are stored under the best possible conditions.

6.3 Biocide use

The use of effective, broad-spectrum biocides is the key to prevention of microbiological spoilage.

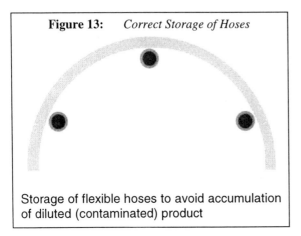

Figure 13:   *Correct Storage of Hoses*

Storage of flexible hoses to avoid accumulation
of diluted (contaminated) product

All aqueous solutions, intermediate products and final formulations should be well
protected. Many problems can be avoided if care is taken to ensure that, wherever
possible, the biocide is the first raw material to be added to the water at the outset of
production. In this way the water is sterilised and raw materials, even if they have a low
level of infection, do not become problematic.

7   BIOCIDES FOR EMULSION PAINT PRESERVATION

Biocides must have a number of essential properties if they are to be successfully used in
emulsion paint manufacture (Table 7).

Table 7:   *Essential Properties of Emulsion Paint Biocides*

❑   Broad spectrum antimicrobial activity
❑   Stable over a wide pH range
❑   Temperature stable
❑   Water soluble at "use concentrations"
❑   Compatible with wide range of paint types/raw materials
❑   No effect on colour
❑   No effect on rheology
❑   Acceptable toxicology/ecotoxicology
❑   Cost effective

The wide range of micro-organisms found in emulsion paints dictates the use of broad
spectrum products. However, many biocides are less active against fungi than bacteria (or
vice versa) and it is therefore important to ensure that biocide manufacturers' claims are
backed up by results of realistic in-use tests.

Most emulsion paints are alkaline in nature, although few will exceed pH 9.5 and some
may be slightly acidic. Thus stability over the range pH 5 to 9.5 is a necessary property.

Temperatures of up to 50°C may occur for short periods during production and the
finished paint may be stored at over 40°C depending on the climate. So stability for
prolonged periods at these elevated temperatures is an important prerequisite.

Micro-organisms inhabit the aqueous phase of an emulsion and thus it is necessary for the biocide not only to be water soluble but also to remain in the aqueous phase.

Compatibility is a difficult property to define. It will include effects on colour and rheology, but lack of this property may manifest as coagulation, loss of corrosion inhibition, deactivation of defoamers, etc.

Some biocides may cause yellowing and others may combine with other raw materials, especially metal salts, to give pigmented complexes.

Any change in rheology brought about by addition of a biocide will be unacceptable unless the formulation can be so designed to take such an effect into account However, serious changes in viscosity which may occur, especially with some acrylic formulations, will render a biocide unusable.

Increased concern about workforce and consumer safety and having a greater regard for the environmental impact of products has led to demands for more toxicologically/ecotoxicologically acceptable biocides. Unlike other raw materials, biocides must be active against living organisms in order to fulfil their function. What is important to consider is the necessary balance between microbiological activity and the effects of the product on higher organisms.

Even if all of the above properties are fulfilled, the ideal biocide will not be used if it is prohibitively expensive. The ratio of required dosage to price (its cost effectiveness) is, therefore, crucial to the success of such a product.

8   MIT/BIT A UNIQUE ISOTHIAZOLINONE COMBINATION BIOCIDE

8.1 Isothiazolinone Biocides

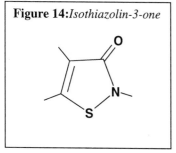

Figure 14:*Isothiazolin-3-one*

Although no single active agent meets all the requirements of the ideal wet-state paint antimicrobial, one biocide group stands out as being the "best we have" and those are products based on the isothiazolin-3-one structure (Figure 14).

The combination of 5-chloro-2-methyl-4-isothiazolin-3-one (CIT) and 2-methyl-4-isothiazolin-3-one (MIT) was described by Elsom in 1988 as probably the most cost effective biocide for industrial preservation. This combined product, along with the benzyl derivative (1,2-benzisothiazolin-3-one - BIT) either on their own or combined with other actives comprise the majority of modern wet-state biocides currently in use.

In the CIT/MIT blend the chlorinated species has by far the greater biocidal efficacy but is also the less stable of the two components. Although MIT alone has relatively low antimicrobial performance it has recently been discovered that this compound has truly synergistic activity when combined with BIT and such a blend has several advantages over CIT/MIT products.

The combination of MIT/BIT blended at equal active agent concentrations derives its activity in three main ways.

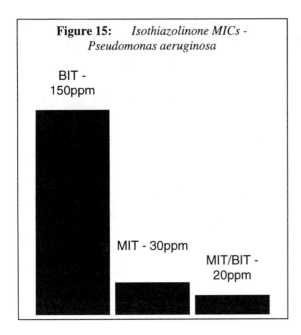

Figure 15:   *Isothiazolinone MICs -*
Pseudomonas aeruginosa

BIT - 150ppm

MIT - 30ppm

MIT/BIT - 20ppm

Firstly, it is intrinsically more active than either compound alone (Figure 15). Secondly, it has a much wider activity spectrum than the two single actives (Figure 16) and thirdly, by its enhanced stability when compared with CIT (Figure 17), required initial addition levels are often similar to those of CIT/MIT blended products.

The MIT/BIT based biocide has a number of important advantages over both CIT/MIT and BIT alone and these are summarised in Table 8.

Figure 16:   *MIC Values of BIT, MIT and MIT/BIT Against Bacteria and Fungi*

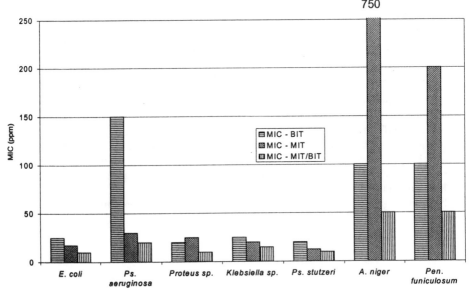

Figure 17: *Relative Stability of CIT and MIT/BIT in an Aggressive Formulation*

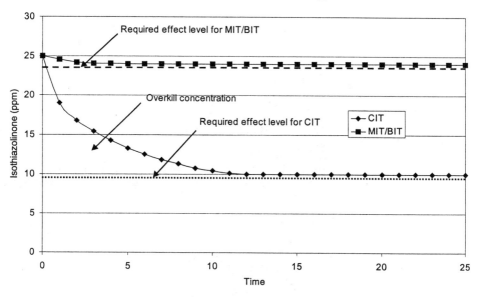

Table 8: *Advantages of MIT/BIT Combination Biocide*

❑ AOX free	❑ Low salts content
❑ Solvent and VOC free	❑ Lower toxicity compared with CIT/MIT and BIT at use concentrations
❑ Unaffected by proposed labelling restrictions on CIT/MIT	
❑ HCHO/aldehyde free	❑ Broader microbiological activity spectrum than BIT
❑ pH stable at > 9	❑ Heavy metal free
❑ Active substances BgVV and EU Food Contact Additive approved	❑ Nitrate free
❑ Temperature stable vs. CIT	❑ Biodegradable/non-persistent
❑ Reducing agent stable	❑ Bivalent metal ion free
❑ Emulsifier free	❑ Truly synergistic activity
❑ Lower skin sensitisation potential	❑ Long lasting efficacy

The long lasting efficacy of a 2.5/2.5% MIT/BIT formulated product is demonstrated by multi-challenge testing carried out in two water-based paints in which its activity was compared with that of a 1.5% CIT/MIT formulation (Figure 18.).

Figure 18: *Comparison of MIT/BIT and CIT/MIT in Two Paints*

Test Sample	Number of Challenges Surviving (' = okay)												
	0	**1**	**2**	**3**	**4**	**5**	**6**	**7**	**8**	**9**	**10**	**11**	**12**
Paint 1 - Blank	█												
+ 0.05% **CIT/MIT Biocide**	█	█											
+ 0.10% **CIT/MIT Biocide**	█	█	█										
+ 0.15% **CIT/MIT Biocide**	█	█	█	█									
+ 0.20% **CIT/MIT Biocide**	█	█	█	█									
+ 0.05% **MIT/BIT Biocide**	█	█											
+ 0.10% **MIT/BIT Biocide**	█	█	█										
+ 0.15% **MIT/BIT Biocide**	█	█	█	█	█	█	█	█	█	█	█	█	█
+ 0.20% **MIT/BIT Biocide**	█	█	█	█	█	█	█	█	█	█	█	█	█
Paint 2 - Blank	█												
+ 0.05% **CIT/MIT Biocide**	█	█											
+ 0.10% **CIT/MIT Biocide**	█	█	█										
+ 0.15% **CIT/MIT Biocide**	█	█	█										
+ 0.20% **CIT/MIT Biocide**	█	█	█										
+ 0.05% **MIT/BIT Biocide**	█	█	█	█									
+ 0.10% **MIT/BIT Biocide**	█	█	█	█	█	█	█	█	█	█	█	█	█
+ 0.15% **MIT/BIT Biocide**	█	█	█	█	█	█	█	█	█	█	█	█	█
+ 0.20% **MIT/BIT Biocide**	█	█	█	█	█	█	█	█	█	█	█	█	█

Paint 1 - pH 9.44, Paint 2 - pH 9.18, samples tested after storage of biocides/paints at 40°C for four weeks

9 CONCLUSION

A wide range of bacteria, moulds and yeasts can infect and cause the deterioration of water based paints and they may originate from a number of sources. However, if care is taken to ensure the good quality of raw materials and make-up water and, if good manufacturing processes and plant hygiene are employed, many microbiological problems seen in manufacture and afterwards may he avoided. Such procedures, allied with the use of effective broad-spectrum biocides, will enable long-term microbiologically trouble-free, production to take place.

Although the "ideal biocide" for the wet-state protection of paints still waits to be discovered the recent development of the methyl/benz-isothiazolinone blend offers many advantages over currently used preservatives.

References

I.   Briggs, M. A., *Paint Research Association Technical Report TR14177, 1977.* Paint Research Association, Teddington, U. K.

2. Briggs, M. A., "Emulsion Paint Preservation - Factory Practice and Hygiene," *Paint Research Association Technical Report TR18178,* May 1980. Paint Research Association, Teddington. U.K.

3. Huddart, G., "In-Can Preservatives for Emulsion Paints," *Unpublished,* February 1983.

4. Jakubowski, J. A., Gyuris, J. and Simpson, S. L., "Microbiology of Modern Coatings Systems," *JCT* 1983, 55 (707) 49.

5. Miller, W. G., "Incidence of Microbial Contamination of Emulsion Paints During the Manufacturing Process," *JOCCA,* 1973, 56 (7) 307.

6. Opperman, R. A. and Goll, M., "Presence and Effects of Anaerobic Bacteria in Water-Based Paint," *JCT,* 1984, 56(712) 51.

7. Skinner, C. E., *Paint Oil Colour Journal,* 1970, 157. 177.

8. Woods, W. B., "Prevention of the Microbial Spoilage of Latex Paint," *J. Waterborne Coatings,* Nov. 1982.

THE ENHANCED PERFORMANCE OF BIOCIDAL ADDITIVES IN PAINTS AND COATINGS

M. Edge, Ken Seal*, N. S. Allen, D. Turner and J. Robinson

Department of Chemistry and Materials
the Manchester Metropolitan University
Chester Street, Manchester M1 5GD, U.K.

*Thor Specialities U.K., Earl Road, Cheadle Hume, Cheshire SK8 6QP

1 INTRODUCTION

Surface coatings are applied to constructional materials to enhance performance and prolong life. A shift to waterborne coatings has rendered the chemical composition of these protective materials particularly susceptible to microbial attack. Microbial spoilage of paint films and coatings has been estimated to cost £15 billion per annum in Europe alone. To be effective, biocides must reside at or near the upper surface of the coating. When the coating is exposed to the natural environment, biocides added in their free state are prone to aqueous extraction and degradation (thermal and UV fragmentation). To overcome such losses formulators find it necessary to include relatively high initial loadings of the biocide.

Biocides are by their nature intrinsically toxic, in this respect any adventitious release to the environment requires an assessment of the relative risk posed. The 5th Environmental Action Plan of the EU is committed to a substantial reduction in the use of biocides. In particular, the Biocidal Products Directive (98/08/EC) is concerned with controlling biocidal products in the market place. Compliance with this directive is required from all member states by 14th May 2000. In this context, a strategy to control the release of biocides is timely, if continued protection is to be afforded to industry and consumer alike. One approach to controlling the release of biocide is to encapsulate in an inert inorganic framework, prior to incorporation in the coating. [1]

2 METHODOLOGY, RESULTS AND DISCUSSION

2.1 Choice of Biocide

Because the objective of this work was to assess the benefits of encapsulation, it was decided to opt for one structural class of biocide. The biocides selected for this purpose were based on the isothiazolinone structure. [2] Examples include: 2-octyl-4-isothiazolin-3-one (OIT), 4,5-dichloro-2-octyl-4-isothiazolin-3-one (DCOIT), 5-chloro-2-methyl-4-isothiazolin-3-one (CIT) and 2-methyl-4-isothiazolin-3-one (MIT) (Figure 1).

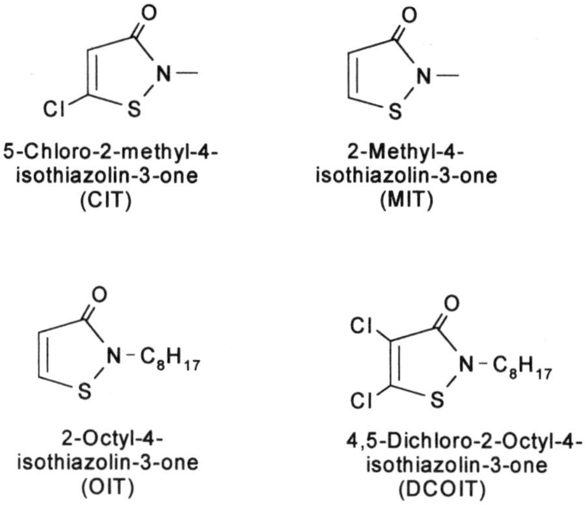

5-Chloro-2-methyl-4-
isothiazolin-3-one
(CIT)

2-Methyl-4-
isothiazolin-3-one
(MIT)

2-Octyl-4-
isothiazolin-3-one
(OIT)

4,5-Dichloro-2-Octyl-4-
isothiazolin-3-one
(DCOIT)

Figure 1   Structural representations of typical isothiazolinone biocides

The literature suggests that the environmental degradation of isothiazolinones proceeds by ring opening through the nitrogen-sulphur bond. Progressive hydrolysis and oxidation produces elemental sulphur, methylamine hydrochloride, malonic acid, and eventually leads to the formation of short-chain carboxylic acids and carbon dioxide. Kazeminski *et al.* have monitored the aqueous degradation of isothiazolinones as a function of temperature and pH [3,4]. They found that the half-life for the first-order decomposition of isothiazolinones was 4.6 days at 40 °C and 0.48 days at 60 °C. As a function of pH at 24 °C the half-lives were 46 days at pH = 8.5 and 3.41 days at pH = 9.62. That is, these biocides are prone to decomposition under basic conditions, in particular via nucleophilic attack by amines and hydroxyl ions.

2.2 Choice of Inorganic Framework

Many inorganic oxides can be manufactured to provide granular, porous materials with high surface areas, which can readily adsorb organic liquids. Preliminary screening of a range of oxides, namely aluminium oxides, titanium dioxides, zinc oxide, hydrotalcites, zeolites and silicas, indicated that the latter two materials were able to retain the largest quantities of biocide.

The surface structure of silica is terminated by silanol sites, which may be classified as vicinal, geminal, isolated (Figure 2). It is the number and distribution of silanol sites that define the adsorption characteristics of a given silica surface [5-7]. Heating the silica surface results in surface dehydration, a process termed 'calcination'. At temperatures below 300 °C, there is a progressive loss of physisorbed water. Above 300 °C only silanol and siloxane bridges remain. Between 400 and 600 °C vicinal silanol groups condense to leave only isolated (and geminal) silanol groups. Between 600 and 800 °C neighbouring silanols condense to leave fewer silanol groups that possess greater acid strength, and stable siloxane bridges. The siloxane bridges are non-polar and so calcination increases the hydrophobicity of the silica surface. Hence, controlled calcination provides a means to control the nature and availability of adsorption sites.

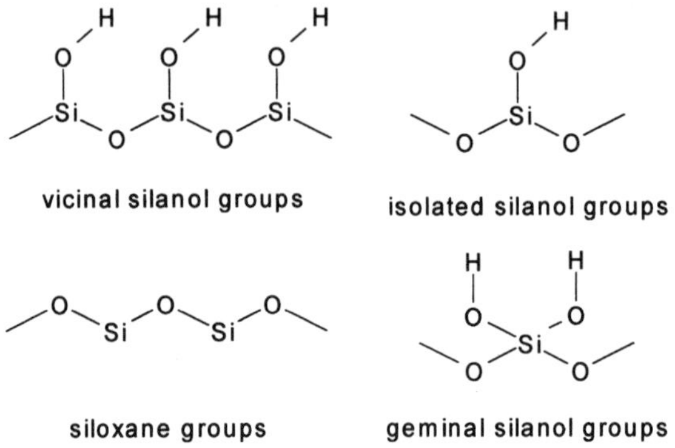

Figure 2   Functional groups present on the surface of porous, amorphous silica

Similarly in zeolites the nature and distribution of acidic surface sites are the properties that affect both activity and selectivity [8-10]. Zeolites are aluminosilicates, i.e. they consist of a framework of $[SiO_4]^{4-}$ tetrahedra in which some of the silicon atoms are replaced by $[AlO_4]^{5-}$, with other cations present (e.g. Na^+, H^+) to balance charges. The tetrahedra are linked by corner-sharing forming bent oxygen bridges, frequently six tetrahedra are connected in a ring (6-ring). Many zeolite structures are based on a secondary building unit that consists of 24 silica-alumina tetrahedra linked together to form a 'basket-like' structure referred to as a sodalite unit (or β-cage). Zeolite-Y consists of the sodalite units linked together by oxygen bridges between four of the eight 6-rings in a tetrahedral array, forming hexagonal prisms.

Depending on its hydrothermal history zeolite-Y may show both Lewis and/or Bronsted acidity. The latter stems from protons occupying exchangeable cation positions following appropriate pre-treatment to render it acidic. Thermal and hydrothermal treatments cause a marked increase in Lewis acidity at the expense of Bronsted acidity. Zeolites can also be dealuminated. This can be achieved by the steam calcination of ammonium zeolites. Here areas of the zeolite are destroyed to give an amorphous aluminosilicate phase, and then aluminium ions from the zeolite framework are hydrolysed out of the crystal lattice. This leaves a vacancy in the lattice, a 'hydroxyl nest' consisting of four silanol groups. This vacancy can be filled by the migration of silica into the lattice. The silicon reinsertion is probably a vapour phase process aided by the steam atmosphere, effectively healing the structure. Non-framework aluminium in dealuminated-Y is a major source of Lewis acid sites, but it is also believed that hydroxyl condensation reactions convert Bronsted sites to Lewis sites without the loss of further structural integrity of the framework (Figure 3). In addition to modifying acidity, increasing the ratio of Si:Al causes a slight decrease in the size of cavities, a decrease in the number of cations and increased hydrophobicity.

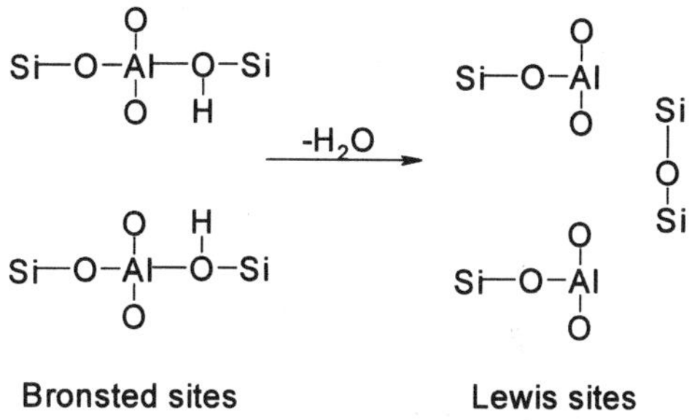

Figure 3 Conversion of acid surface sites in zeolites treated by steam calcination

2.3 The Determination of Structure-Performance Relationships

To assist the optimisation of available adsorption sites, Flow Microcalorimetry was used to better understand the interactions between the biocide and active silanol sites. The flow microcalorimeter (FMC) is able to measure heat exchanged as a result of adsorption/desorption occurring under flow conditions [11]. A differential refractometer, downstream to the calorimeter allows quantification of an adsorbate. The amount of solute adsorbed on the solid is calculated from the comparison of parallel experiments conducted with the solid and a non-adsorbing probe. The difference between the refractometer signals of the two experiments permits the calculation of the amount of adsorption. Appropriate calibration also allows an indirect measurement of the enthalpies associated with adsorption and desorption. The FMC system used consisted of a Microscal$^©$ 3V FMC, with a PTFE fluid path upgrade, linked to a Waters$^®$ 410 differential refractometer (Figure 4). Data was output to a Perkin-Elmer$^©$ Nelson 900 series data station with appropriate software. Experiments were conducted using a cell temperature of 28 °C ± 0.5 °C and adsorbate concentrations of 0.3% w/v. Decahydronaphthalene was used as the non-adsorbing probe, and cyclohexane (HPLC grade dried over 4 Å molecular sieve) as the solvent for the isothiazolinones. The flow rates for both the non-adsorbing probe and the solvent were 3.30cm^3 h^{-1}. Within the FMC cell the adsorbent (volume 0.15 cm^3; 350 – 800 mg) was equilibrated, with cyclohexane, for 18 hours before measurements commenced. This conditioning period was necessary to remove the vast majority of loosely bound surface water from the silica substrate. Initially the adsorbate in cyclohexane was passed through the cell containing the siliceous adsorbent, and after adsorption had taken place cyclohexane was passed through the cell to facilitate desorption. For both adsorption and desorption measurements, data reported is that for the average result between two runs.

To probe interactions between active silanol sites and the isothiazolin-based biocides a number of model probes were investigated [12]. The adsorbates (1-methylpyrrolidin-2-one, pyridine, pyrrolidine, pyrrole, 2-methylthiophene, 2-octyl-4-isothiazolin-3-one, 4,5-dichloro-2-octyl-4-isothiazolin-3-one and 2-cyclopenen-1-one,) varied in basicity, polarity and π-character. The amounts of the adsorbates retained by the silica were determined (Figure 5), along with enthalpy of adsorption (ranging from -5.5 kJ mol^{-1} to -57.8 kJ mol^{-1}) and enthalpy of desorption (ranging from 5.6 kJ mol^{-1} to 26.1 kJ mol^{-1}). For the majority of the adsorbates the enthalpy of adsorption is consistent with hydrogen bonding to isolated silanols (Figure 6).

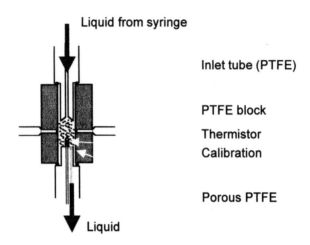

Figure 4   Schematic of Flow Microcalorimeter Cell

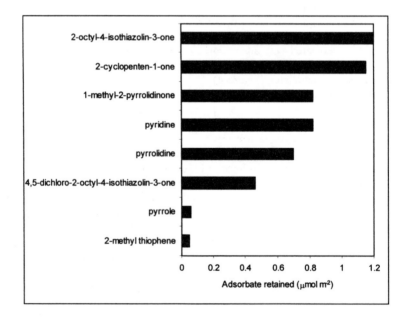

Figure 5   The amount of model probe retained (μmol m^{-2}) by silica sample 2.

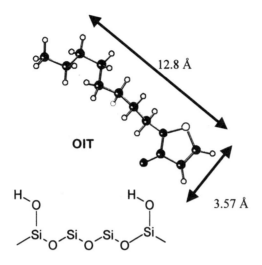

Figure 6 Schematic of hydrogen-bonding between the biocide OIT and isolated silanol groups

Although increasing basicity enhanced the adsorption enthalpy and hence the strength of associations, desorption was inhibited when a carbonyl, or unsaturated carbonyl, group was adjacent to the active basic centre. Bulky electron-withdrawing agents (chlorine atoms) substituted at the double bond of the unsaturated carbonyl reduced the adsorption considerably. Using a simple model assuming isosteric adsorption, and with a knowledge of specific surface area of the carriers, the approximate distance between adsorbate (biocide) molecules on the surface of the siliceous frameworks was determined. This enabled the percentage of active sites of various acid strengths to be determined as a function of calcination temperature. This work facilitated a better understanding of how the surface chemistry of the carrier could be 'tailor-made' to retain a specific biocide structure.

On this basis the porosity and surface composition of a number of silicas and zeolites were varied systematically to maximize retention of the isothizolinone structures. For the sake of clarity, data is represented here for only four silicas (Table 1) and three zeolites (Table 2). Silicas 1 and 3 differ in their pore dimensions, these being *ca.* 20 Å and 180 Å respectively. Silicas 2 and 4, their counterparts, have been calcined to optimise the number and distribution of isolated silanol sites. Zeolites 1 and 2 are the Na- and H- forms of zeolite-Y respectively. Zeolite 3 is the H-Y zeolite after subjecting to steam calcination, thereby substantially increasing the proportion of Si:Al in the structure. The minimum pore dimensions of these materials were around 15 Å, selected on the basis that energy-minimized structures obtained by molecular modelling predict the widest dimension of the bulkiest biocide (OIT) to be *ca.* 13 Å, thereby allowing entry to the pore network.

Table 1   Porosymmetry and Particle Size Data for Porous Silicas*

Silica Sample	B.E.T. Surface Area $(m^2 g^{-1})$	Pore Volume $(ml\ g^{-1})$	Average Pore Diameter (Å)	Average Particle Size (μm) d_{50}
1	728	0.41	23	4.6
2	562	0.34	25	4.3
3	395	1.78	180	6.2
4	368	1.68	182	6.5

Table 2   Porosymmetry and Particle Size Data for Zeolites*

Zeolite Sample	B.E.T. Surface Area $(m^2 g^{-1})$	Pore Volume $(ml\ g^{-1})$	Average Pore Diameter (Å)	Average Particle Size (μm) d_{50}
1	706	0.38	15	3.8
2	581	0.36	17	4.4
3	602	0.39	25	5.1

Particle size was measured using a Malvern Mastersizer, and nitrogen porosymmetry undertaken using a Sorpty 1750 instrument.

2.4 The Relative Retention of Biocide after Aqueous Leaching

A Retention Factor, R, was defined to distinguish between the retention of biocide by weight and its relative potency. The potency of the biocide was measured in terms of the Minimum Inhibitory Concentration (MIC) necessary prevent the growth of the microorganism, *Aureobasidium pollulens* (for the biocide OIT the MIC is 36 mg/l). From this work potential carriers were defined as those having a value of R in excess of 0.6. This limit excluded all materials, excepting those porous silicas and zeolites with pore areas in the pore size range 20 to 50Å of greater than 35 m^2g^{-1}. Again for the sake of clarity data is presented only for the retention of OIT. Figures 4 and 5 illustrate the relative retention of OIT by the silica and zeolite carriers described in Tables 1 and 2, after fixed quantities, loaded with 0.3 g biocide per 1.0 g silica, were placed in a litre of water for 90 minutes with continuous stirring.

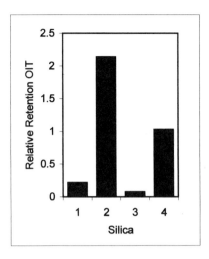

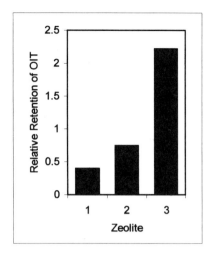

Figure 7 The relative retention of OIT by the porous silicas Described in Table 1.

Figure 8 The relative retention of OIT by the zeolites described in Table 2.

Figure 7 demonstrates the benefit of calcination, to control the number and distribution of silanol sites, in order to promote acid-base interactions between the silica surface and isothiazolinone biocide. Compared with silica samples 1 and 2, the calcined silicas (samples 3 and 4) show a ten-fold increase in the relative retention of the biocide OIT. In addition, the egress of the biocide is restricted as the pore dimensions are reduced. Comparing silicas 2 and 4, where the relative retention of OIT is approximately doubled as the pore dimension is decreased from 180Å to 20Å, the influence of steric hindrance is obvious. Figure 8 illustrates the benefit of increasing the proportion of Si:Al in a zeolite structure on the relative retention of the biocide OIT. Again this supports the premise that increasing the relative acidity of available silanol sites, coupled with the increased hydrophobicity of the porous structure, increases the relative retention of the isothiazolinones, when subjected to aqueous extraction.

Figure 9 illustrates the fact that the release of biocide from the carriers is a dynamic process. Here a quantity of loaded carrier was slurried with a fixed volume of water and aliquots taken after 1 hour. From previous experiments it was found that after an initial period of rapid release, a steady-state concentration of free biocide was present in the aqueous extract. To probe the effects of repetitive extraction, the carrier was filtered from the slurry, the water replenished and the process repeated. It can be seen that only after ten successive extractions does the amount of the biocide OIT released by the carrier fall below the MIC. It should be noted that the conditions employed to illustrate this continuous release are rather more severe than would be experienced when the loaded carrier is incorporated in a coating (see Section 2.5).

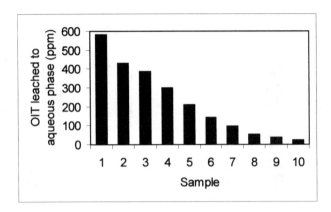

Figure 9   OIT (ppm) leached to an aqueous phase following repetitive extractions
(1 to 10) from Silica Sample 2

2.5 The Performance of the Encapsulated Biocide

The appropriate amount of biocide carrier composition needed to yield a specified dry film concentration of biocide was added to a fixed quantity of waterborne acrylic paint and premixed using a high-speed disperser. This dispersion was then transferred to a Silverson-type mixer to obtain a finer dispersion. At this stage the influence of addition of silica or zeolite on the physical/mechanical properties of the paint film was assessed, though only minor changes in properties were noted, which could be eliminated by appropriate adjustments to the paint formulation.

The biocidal efficacy in the afore mentioned paint formulations was determined by a rapid screening test to determine the zones of inhibition around cured, painted discs placed face down on a solid agar surface seeded with the fungus *Cladosporium cladosporioides*. The results of this study provided sufficient evidence to support the observation that paint films containing the loaded carriers retained more biocidal activity after leaching than those containing free biocides (Figure 10). Following this long-term fungicidal activity was assessed according to BS3900: Part G6 (with *Alternaria alternata* included in the standard inoculum). Here masterboard panels were brush coated on one side with the test paints, and the coatings (two) cured at 60°C for two days. The coated panels were weathered in a QUV apparatus under the following conditions: 125 hours exposure to UVA (340 nm) at 40°C with water spray for 1-hour duration, regularly at 24-hour intervals. After 2 months exposure the blank (no biocide) & free biocide showed 85% fungal growth c.f. 32% and 25% fungal growth on silica and zeolite carried biocide (both 300 ppm) respectively.

Figure 10   Effect of carrier after 24 hours aqueous leaching on emulsion paint containing 1200 ppm OIT: without carrier, with zeolite and with silica (left to right).

2.6 Environmental Benefits

As mentioned in Section 2.1, isothiazolinone biocides are stable only in the pH range 4-10, yet modern paint formulations can have pHs as high as pH=10-12. Furthermore isothiazolinones degrade on storage above *ca.* 40°C and on exposure to UV radiation. The purpose of this part of the study was to demonstrate any increased stability afforded to the biocide when contained within the porous siliceous framework. Here loaded biocides and stock biocide were exposed to UV-visible radiation (>300nm) using a 500W high pressure Hg/W lamp at 50°C and relative humidity 50%. Samples were withdrawn at regular intervals and the amount of biocide remaining determined by various spectroscopic methods. The results indicate that carried biocide is afforded protection to both thermal and UV degradation. Again for the sake of clarity, this is illustrated only for the UV exposure of OIT in Figure 11.

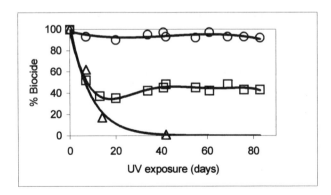

Figure 11   The reduction in concentration of the biocide OIT (%) resulting from exposure to UV-visible radiation for free biocide ($\triangle$), biocide loaded on Silica Sample 2 (O) and biocide loaded on Zeolite Sample 3 ($\square$).

3 CONCLUSION

This study has shown that typical coating biocides can be encapsulated within modified silica frameworks. These porous frameworks offer a means to inhibit the aqueous extraction of the biocide. In such combinations the biocides retain their anti-microbial properties, while controlled delivery facilitates a dynamic equilibrium to maintain a minimum inhibitory concentration at the coating interface, over an extended time period. There is evidence that biocide housed in such frameworks has a longer effective activity for a given initial concentration, since it is to some extent protected from the usual environmental degradation processes.

References

1. UK Patent Application No PCT/GB 99/02796, "*Particulate Carriers for Biocide Formulations*".
2. J. Gillatt, *Polymers Paint Colour J.*, 1993, May 12.
3. S. F. Kazeminski, C. K. Brackett and J. D. Fisher, *J. Agric. Food Chem.*, 1975, 23, 1060 – 1068.
4. S. F. Kazeminski, C. K. Brackett, J. D. Fisher and J. F. Spinnler, *J. Agric. Food Chem.*, 1975, 23, 1068 – 1075.
5. R. K. Iler, *The Chemistry of Silica*, Wiley, New York, 1979.
6. L. T. Zhuravlev, *Pure Applied Chem.*, 1989, 61, 1969.
7. L. T. Zhuravlev, *Langmuir*, 1987, 3, 316.
8. B. C. Gates, "*Catalytic Chemistry*", Wiley & Sons, New York, 1992.
9. J. W. Ward, *J. Catal.*, 1968, 11, 251.
10. H. Pfeifer, D. Freude, and M. Hunger, *Zeolites*, 1985, 5, 274.
11. D.P. Ashton, R.N. Rothon, in "*Controlled Interfaces in Polymeric Materials*", H. Ishida (Ed.) Elsevier Science, New York, 1990.
12. M. Edge, N. S. Allen, D. Turner, J. Robinson, *J. Mat. Sci.*, accepted March 2000.

Acknowledgements

We gratefully acknowledge funding for this work from the DTI and EPSRC
(D. Turner and J. Robinson), through the LINK Surface Engineering Programme.
The work is collaborative with contributions from a consortium comprising *the* Manchester Metropolitan University, Crosfield U.K., Thor Specialities U.K., Liquid Plastics and the Paint Research Association.

Disinfection

Actual situation in developing a European Standard for testing of surface disinfectants

Dr. Gerhart Heinz

Consultant
Auf der Eck 15a, D-76547 Sinzheim, Germany
Tel. +49 7221 82640
E-Mail: Gerhart.Heinz@t-online.de

1 INTRODUCTION

Up to now there is a strange situation in the area of disinfection: A disinfectant judged to be effective in a concentration A in France might be effective in the UK in a higher concentration B or even in another concentration C in Germany.

The reason for this confusing situation is that all efficacy data are depending from the test method used to find the effective time - concentration ratio of the disinfectant.

2 DEVELOPMENT OF EUROPEAN STANDARDS FOR CHEMICAL DISINFECTANTS

1. European test method for chemical disinfectants

In 1989 the CEN/TC 216 (figure 1) was astablished to produce harmonized European test methods for disinfectants and antiseptics used in medicine, veterinary and food hygiene.

Figure 1

Tasks of the CEN / TC 216 Project

"Chemical disinfectants and antiseptics"
- Standardisation of terminology, requirements and test methods including the potential efficacy under conditions of use.

- Recommendations for use and labelling in the total area of chemical disinfection and antiseptics.

- Activities include the areas of agriculture (except pesticides), household, food hygiene and other industrial, commercial, medical and veterinary applications.

To harmonize the work of the 3 different Working Groups (Human medicine, Veterinary and Food hygiene) the Horizontal Working Group was installed (Figure 2).

Figure 2

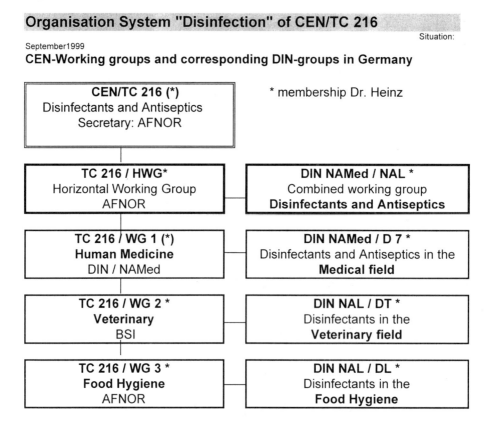

Organisation System "Disinfection" of CEN/TC 216

Situation:
September1999
CEN-Working groups and corresponding DIN-groups in Germany

CEN/TC 216 (*) Disinfectants and Antiseptics Secretary: AFNOR	* membership Dr. Heinz

TC 216 / HWG* Horizontal Working Group AFNOR	**DIN NAMed / NAL *** Combined working group **Disinfectants and Antiseptics**
TC 216 / WG 1 (*) **Human Medicine** DIN / NAMed	**DIN NAMed / D 7 *** Disinfectants and Antiseptics in the **Medical field**
TC 216 / WG 2 * **Veterinary** BSI	**DIN NAL / DT *** Disinfectants in the **Veterinary field**
TC 216 / WG 3 * **Food Hygiene** AFNOR	**DIN NAL / DL *** Disinfectants in the **Food Hygiene**

2. SURFACE DISINFECTION

Whithin the Horizontal Working Group (HWG) a Task Group about Surface Disinfection was built by experts from France, UK, Belgium and Germany, the „Surface Task Group (STG).
The place where the STG in located in the project tasks of CEN / TC 216 is shown in figure 3.

Figure 3

Project tasks CEN / TC 216

Development of standard test methods for estimating chemical disinfectants and antiseptics

Phase 1	**Basic activity** (Screening test)	- bactericidal activity - fungicidal activity - sporocidal activity - virucidal activity
	Working group:	- HWG
Phase 2	**Laboratory test methods simulating practical conditions**	
	step 1: special requirements:	- WG 1 - WG 2 - WG 3
	step 2: carrier test for special fields of application:	- WG 1 **Surface Task** - WG 2 **Group** - WG 3 **(STG)**
	skin- / hand- disinfection tests	
Phase 3	**Field tests on naturally contaminated surfaces**	
	if necessary and possible	

The aim of the **STG** is to develope a general surface test which will be modified by the 3 working groups for their specific requirements.

The first draft (finished in 1993 and revised in 1995) describes a test method for determination of the bactericidal activity of a chemical disinfectant applied to a contaminated surface under laboratory conditions:

Stainless steel disks are contaminated with a bacterial test suspension and dried. The disinfectant is applied on the dried film on the disk and kept at a specified temperature for a defined time. The disk is than transferred to a previously validated neutralization medium to stop the action of the disinfectant. The cfu of surviving bacteria recovered from the surface is determined quantitatively.

In a parallel test with water instead of disinfectant the colony forming units (cfu) of surviving bacteria are determined and the reduction in viable counts is calculated.

The 3 working groups discussed the method and all the groups asked for a variant including mechanical action. WG2 also considered to use a wooden carrier instead of stainless steel.

In the following meetings the STG discussed different variants with mechanical action.

The first mechanical variant uses a presoaked cotton swab on the surface covered with the disinfectant solution, the second variant uses a presoaked swab without disinfectant on the test surface. Both methods have been tested by Prof. Koller in Vienna and the first variant showed a higher efficacy than the second (using more disinfectant on the surface).

It was decided to incorporate both variants in the basic methodology.

In the last meeting of the STG in June 1999 the ratio between microorganisms and disinfection liquid was discussed and the suggestion was to use 15 ml/m².

The method will now be tested in a ringtest by 8 dfferent laboratories in Europe. A preliminary range finding will be done in Vienna in October 1999. Test oganisms will be *Stapylococus aureus* and *Pseudomonas aeruginosa*, exposure time will be 5 minutes. The test surfaces will be stainless steel disks (2 cm diameter) with a specific finish. The problem of a standardized drying procedure was discussed. There was no decision about the number of replicates of the test.

The method will now be tested in a ringtest by 8 different laboratories in Europe. When the results of the ringtest are aviable they will be given to the 3 working groups for discussion.

3. LABORATORY CONDITIONS

From the experience in performing surface tests it is known that room temperature and humidity have a strong influence on the test results, so only in a laboratory whith standardized temperature and humidity the tests will be reproducible.

Since only a few microbiological laboratory has the equipment to standardize these parameters it is not intended to fix these parameters in the standard.

4. CONCLUSIONS

In a future European Standard for Surface disinfection not all details for achieving reproducible test results will be described. So every laboratory which intends to do the tests has to find by ist own experience how to get reproducible test results. This means you have to standardise the Standard.

Plastics

MICROBIOCIDES FOR PVC AND OTHER POLYMERS

R. Borgmann-Strahsen

Akzo Nobel Chemicals
Chemicals Research Düren
Germany

1 INTRODUCTION

There are two main scopes for the use of microbiocides in PVC and other polymers. The classical approach is the protection of susceptible plastic material whereas a more recent approach is the achievement of biocidal surfaces.

1.1 Protection of material

It is a well-known fact that specific plastic materials like flexible PVC, Polyurethane or Silicone may be easily attacked by microorganisms leading to discoloration or mechanical failures.[1-4] This susceptibility to microbial attack is mainly attributed to the plasticiser content of the material as well as other ingredients such as stabiliser or antioxidants.[5,6] The predominant organisms on the surface of those plastics are fungi and actinomycetes and it is said that by the action of their extracellular enzymes other organisms such as bacteria may be able to grow on the material.[7]

Plastic materials may be protected against microbial attack by incorporation of an active fungistat.[8] These are substances that lead to a suppression of fungal growth on the plastic materials.

1.2 Biocidal Surfaces

In recent years, increasing public concern about food-borne pathogens and other germs in the environment has been driving consumer demand for antimicrobials in products made of polyolefins and other plastics ranging from household goods such as cutting boards to children´s toys.[9] Also in industrial, institutional and medical areas such biocidal surfaces are of growing interest.[10]

Microorganisms that might cause problems due to their presence on specific surfaces are mainly pathogenic bacteria rather than fungal organisms. Consequently, a biocide for such an application would have to be highly active against bacteria. In contrast to the protective application a -static i.e. growth inhibitive action of the incorporated biocide would not be enough. What would be needed would be a -cidal i.e. a killing effect. Due to the fact that mainly bacteria would have to be combated the requested product would have to be a strong bactericide.

2 ANTIMICROBIAL SUBSTANCES

To be applicable in plastics the antimicrobial substances would not only have to be active against the requested microorganisms but they would also have to meet very specific requirements to show the desired effect in the final product:

- heat stable
- (soluble in plasticiser)
- compatible with polymer
- low leaching rate
- UV stable
- available on plastic surface
- environmentally safe.

2.1 Protection of material

Some selected fungistats that are used to protect plastic materials against fungal attack are listed below:
- Intercide ABF: 10, 10'oxibisphenoxarsine (OBPA)
- Intercide OBF: 2-n-octylisotiazolin-3-one (OIT)
- Intercide IBF: 3-iodo-2-propinylbutylcarbamate (IPBC)

2.2 Biocidal Surfaces

There are not yet many products on the market that show bactericidal activity and meet at the same time the requirements for application in plastics:
- Intercide ABF: 10, 10'oxibisphenoxarsine (OBPA)
- Intercide ZNP: Zinc pyrithione (ZNP)
- Triclosan: 2, 4, 4'-trichloro-2'-hydroxy-diphenyl ether

3 MICROBIOLOGICAL TESTS

3.1 Protection of materials

3.1.1 Test Method. There are several standard methods available to investigate the fungistatic activity of plastic materials (e.g. ISO 846, ASTM G21, ASTM E 1428). In the given study a method has been used which is not yet a standard test method but which is at the moment passing through the official procedure to become an ISO standard: The NSA method.

In this method a disk of the test foil is placed on the surface of solidified nutrient salts agar (NSA) in a petri dish. Subsequently, the specimen as well as the surrounding agar is covered with a thin layer of molten NSA containing the mixed spores of five test fungi. After incubation for 21 days at 25 °C the test is evaluated. Due to the fact that the NSA does not contain any carbon source fungal growth on the agar is very limited. If an active fungistat was incorporated in the test foil and had migrated into the agar this is seen as a zone of inhibited growth or inhibited germination. If the foil is susceptible to fungal attack and it is not protected by an active fungistat this will manifest in fungal growth above the specimen being significantly stronger than on the nutrient salts agar. Observation of significant growth means that the NSA test is not passed.

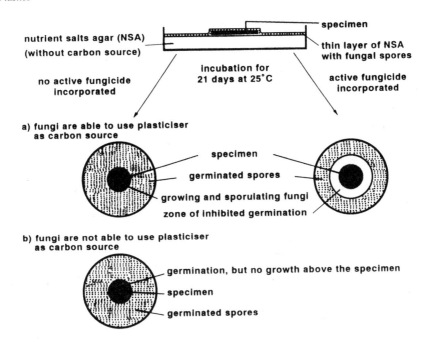

Scheme 1 *NSA Test Procedure*

3.1.2 Test Results. The three biocides as listed under 2.1 were incorporated in calandered flexible PVC foils which were investigated according to the NSA test. Table 1 gives an overview on the test results.

Active ingredient	% in the foil	Zone of inhibited growth [mm]	Fungal growth upon the specimen*
OBPA	0,05	26	-
IPBC	0,2	9	-
IPBC	0,1	5	-
OIT	0,2	13	-
OIT	0,1	5	-
-	-	0	++

*: -:   no fungal growth
   (+):   initial growth comparable to that on NSA
   +:   slight fungal growth
   ++:   strong fungal growth

Table 1 *NSA Test Results on Calandered PVC Films*

All three biocides tested showed distinct fungistatic activity in the NSA test and consequently adequate protection against fungal attack. For practice application potential weathering of the foils e.g. by water leaching should be taken into consideration. In the given presentation it is not possible to go into detail regarding weathering but results of a more detailed study including leaching outcomes have been published.[12]
To demonstrate the fungistatic activity in another type of plastic polyurethane foams, equipped with different concentrations of OIT, have been investigated. Results of the NSA test on those samples are listed in table 2.

Active ingredient	% in the foam	Zone of inhibited growth [mm]	Fungal growth upon the specimen*
OIT	0,14	0	(+)
OIT	0,24	7	-
-	-	0	+

*: -:   no fungal growth
(+): initial growth comparable to that on NSA
+:   slight fungal growth
++:   strong fungal growth

Table 2 *NSA Test Results on PU Foams*

It is obvious that in the PU foam material higher concentrations of OIT are needed to achieve adequate protection than in the flexible PVC foils although the latter material as such seems to be more susceptible to fungal attack. However, strong fungistatic activity leading to an inhibition zone in the NSA test is important in the PU foam area. Those materials are often used for bedding or upholstery. In this application loading of the foam material with sweat and other organic materials during use is normal. Fungal organisms may grow on this organic soiling and may serve as nutrient for dust mites. Fungal spores as well as dust mite excrement's may cause allergic reaction and consequently should be avoided in the bedding and upholstery area. So it is important to use a biocide in this field which does not only protect the material itself but also its surrounding. Protection of the surrounding of a plastic is a property which is usually demonstrated by formation of an inhibition zone in the NSA test or similar microbiological tests.

3.2 Biocidal Surfaces

3.2.1 Test Method. Due to the fact that there does not yet exist an official method for the investigation of the bactericidal activity of surfaces an adequate method had to be developed:
Within small test tubes defined plastic specimen were embedded in a thin agar layer containing the test organisms. The number of organisms was adjusted to different decimal levels ranging from 10^{-1} up to 10^4. Three parallel tubes were tested for each dilution step. After a contact time of 2 and 24 hours at which the tubes were held at 37 °C the test tubes were checked for presence or absence of surviving cells. Via the resulting pattern of tubes with or without growth and the comparison with a blank reference sample the logarithmic reduction of bacterial cells for the single plastic samples was determined. Four different bacteria were included in the test in order to get an impression on the spectrum of activity of the specific biocides: Staphylococcus aureus, Streptococcus faecium, Pseudomonas aeruginosa and Escherichia coli.

A scoring system was fixed to simplify expression of the test results (table 3).

Range of logarithmic reduction	Score
-1 to 0,9	0
1 to 1,9	1
2 to 2,9	2
3 to 3,9	3
4 to 4,9	4
5 to 5.9	5

Table 3 *Scoring System of the Test on Bactericidal Activity*

Active ingredient	% in plastic	Staphylococcus aureus	Streptococcus faecium	Pseudomonas aeruginosa	Escherichia coli
ZNP	0,2 % in LDPE	2	1	0	0,5
ZNP	0,5 % in LDPE	3	0,5	0	0,5
Triclosan	0,2 % in LDPE	2,5	0	0	0,5
Triclosan	0,5 % in LDPE	4	1	0	3
OBPA	0,05 % in PVC	4,5	0	0	5

Table 3 *Results of the Test on Bactericidal Activity after 24 Hours Contact Time*

The given method does fulfil the main requirements for a test on bactericidal activity on surfaces like avoidance of drying of the cells during contact time, equal distribution of the test organisms and easy recovery of each single cell. Due to the fact that the procedure is very time consuming and consequently cost intensive it will be further optimised. However, to get an impression on the activity of different biocides for the given application some results with the given method are listed in table 4. These are the results obtained after 24 hours of contact time. After the shorter contact time of 2 hours there was no significant bactericidal effect detectable.

It was found that the investigated biocides do show indeed a bactericidal effect when incorporated in plastic sheets. It is obvious that this is a relatively slow effect because significant reductions of living cells were not found after 2 hours contact time but after 24 hours. The logarithmic reductions of living cells detected after the latter period of contact time were in most cases significantly below 5. According to the new CEN test procedures for disinfecting agents (e.g. EN 1276) a bactericide has to achieve a logarithmic reduction of 5 within 5 minutes. Taking this into account it is obvious that the bactericidal effect of plastic surfaces cannot be compared with the activity of a disinfectant and consequently it will never be able to replace disinfection in practice - at least not with the products tested in this study.

Another limiting factor of the tested products is the very narrow spectrum of activity which was found to be more or less the same for all three products tested. They all showed by far the best activity against the gram positive bacteria Staphylococcus aureus. In contrast to that no biocidal activity was measured regarding the gram negative bacteria Pseudomonas aeruginosa. Zinc pyrithione and Triclosan both resulted in some activity against Streptococcus faecium (gram +) and Escherichia coli (gram -). OBPA had no effect against S. faecium and a strong effect against E. coli at the given test concentration which was much lower than that of the other products tested.

References

1. K.J. Seal and M. Pantke, *International Biodeterioration*, 1988, **24**, 313
2. H. Becker and H. Gross, *Material und Organismen*, 1974, **9**, 81
3. F. Demmer, *Material und Organismen*, 1968, **3**, 19
4. J.P. Scullin, T.A. Girard and C.F. Koda, *Vinyl Resins*, 1965, **3**, 267
5. K.J. Seal, *Biodeterioration Abstracts 2*, 1988, **4**, 295
6. G. Tirpak, *SPE Journal*, 1970, **26**, 26
7. E. Bessems, *Journal of Vinyl Technology*, 1988, **10**, 3
8. F. Stühlen and E.H. Pommer, *Kunststoffe*, 1983, **73**, 32
9. L. Manolis Sherman, *Plastics Technology*, Febr. 1998, 45
10. A. Hoffmann, *BIOforum*, Sept. 1999, 538
11. R. Borgmann-Strahsen and E. Bessems, *Kunststoffe/plast europe*, *1994*, **84**, 24, 158
12. R. Borgmann-Strahsen and M.T.J. Mellor, *Kunststoffe/plast europe*, 1999, **89**, 17, 68

Performance Fluids

THE APPLICATION OF BIOCIDES IN METALWORKING FLUIDS

M A Wright

Technology Centre
Whitchurch Hill
Pangbourne
Reading RG8 7QR

1 INTRODUCTION

Metalworking fluids are an essential component in the smooth running of many manufacturing industries. The variety of metalworking industries includes metal manufacture, transport equipment manufacture, fabricated metal goods and machinery manufacture. The global market for metalworking fluids was estimated to be approximately 4.4 billion litres during 1997. This accounts for approximately 40% of the entire global market for industrial lubricants.

In general terms, metalworking fluids serve three purposes. They are used to lubricate, reducing the friction between the tool and the work piece, to cool increasing the accuracy of the machining operation and to remove the chip generated by the cutting operation. Broadly speaking, there are three types of metalworking fluid:

1.1 Neat Oils

These are delivered ready for use and are designed to be immiscible with water. They are usually blends of mineral oils or a synthetic basestock, for example a synthetic ester, augmented with additives to assist in metalworking processes. They are water free and, under normal use conditions, are not subject to degradation by micro-organisms.

1.2 Emulsifiable Oils

These are usually mineral oils or synthetic basestocks containing a complex mixture of additives including corrosion inhibitors, extreme pressure additives and emulsifiers. They are almost exclusively oil-in-water emulsions, although in rare instances, invert emulsions may be used. They are normally used between 1%–20% emulsions depending on the application. As these products are water extendible, they are subject to attack by micro-organisms. As a consequence, they are often formulated with one or more preservatives.

1.3 Water Soluble Synthetic Solutions

These are true chemical solutions and are mixtures of soluble polyglycols (to give lubricity), corrosion inhibitors and water soluble extreme pressure additives. They are subject to attack by micro-organisms and as a consequence, they are often formulated with one or more preservatives.

Water extendible metalworking fluids are typically buffered around pH 9.0. They are complex chemical mixtures whose composition can vary from a few ingredients up to twenty separate components, (See Table 1).

Table 1: *Composition of a Typical Metalworking Fluid*

Component	Examples	Function
Base oil	Mineral oil	Lubrication
	Synthetic ester	Lubrication
	Polyglycol	Lubrication
Surfactant	Fatty acid soaps	Emulsification of product
	Petroleum sulfonates	Emulsification of product
	Ethoxylated alcohols	Emulsification of product
Corrosion inhibitors	Organic acids	Film formation
	Amines	Film formation
Extreme pressure additives	Chlorinated paraffins	Lubricity agents
	Sulfurised esters	Lubricity agents
Biocides	Bactericides	Preservatives
	Fungicides	Preservatives
Metal passivators	Triazole	Protect yellow metals
Couplers	Organic alcohols	Increase solubility of polar molecules in MWF concentrates

2 THE MICROBIOLOGY OF METALWORKING FLUIDS

Water extendible metalworking fluids are susceptible to contamination by micro-organisms, used coolants in particular can be very easily colonised. Metalworking systems are open to the air and are constantly inoculated with micro-organisms. Sources of inoculum include:
a) The air around us
b) The skin and hair of machine operators
c) Make-up water used to prepare the emulsion
d) Extraneous organic matter such as food, cigarette ends, etc.
e) Use of the emulsion as a toilet
f) Residual microbial slimes in machines improperly cleaned out
    Once contaminated, the growth of micro-organisms in the coolant can be very quick. Rapid microbial growth may be encouraged by coolant components such as alkanolamines, soaps and oleic acid amides. In addition, poor system design, eg introduction of dead ends, ineffective swarf and tramp oil management, allowing the coolant to stagnate and failure to follow manufacturers instructions can all lead to the rapid development of microbial infections. Bacteria and fungi are the principle spoilage organisms found in metalworking fluids. However, in some instances, yeast may be found as well, [see Table 2].
    The uncontrolled growth of micro-organisms in metalworking fluids can result in a wide range of problems. These can include:
a) The production of foul odours, often encountered if the coolant is allowed to stagnate. The ubiquitous "Monday morning" smell, after a weekend shutdown is a good indicator of a microbial contamination problem.
b) Unsightly discolouration of the product may occur.
c) Drop in pH may be observed. This can result in the corrosion and staining of the work piece and of the machine itself. A fall in pH may also promote the growth of different microbial populations.

Table 2: *Some Microbial Contaminants Found in Metalworking Fluids*

Bacteria	Fungi	Yeast
Citrobacter freundi	*Cephalosporium sp.*	*Candida sp.*
Desulphovibrio sp.	*Fusarium oxysporum*	
Enterobacter cloacae	*Fusarium solani*	
Escherichia coli		
Klebsiella pneumoniae		
Proteus mirablis		
Pseudomonas aeruginosa		
Alcaligenes xyloxidans		
Burkholderia pickettii		
Clavibacter michiganese		
Enterococcus faecium		
Gordona rubropertinctus		
Methylobacterium mesophilicum		
Methylobacterium radiotolerans		
Nocardia globerula		
Phyllobacterium rubiacearium		
Pseudomonas saccharophilia		
Rhodococcus erythropolis		

d)   In emulsifiable products, emulsion instability may occur as a result of the change in pH or direct microbial attack on the surfactants present in the product. Emulsion instability may result in the pooling of oil and loss of lubrication efficiency.

e)   Micro-organisms, in particular the fungi, may form fibrous mats or slimes which may block pipework and filters (see Figure 1).

f)   Microbial contamination may also reduce coolant life having a direct economic impact on the factory owner. This is manifest in terms of increased maintenance, cleaning and disposal costs.

g)   There may also be potential human health effects of having heavily contaminated metalworking fluids circulating in systems for extended periods of time.

h)   In water contaminated neat oils, microbial contamination may result in the formation of stable emulsions resulting in corrosion of the machine and loss of lubricant performance.

There are many steps that can be taken to reduce the chances of microbial spoilage occurring in a metalworking

Figure 1: A Typical Fungal Coloniser

fluid. These include:

a) Prevent the accumulation of swarf in the system. Swarf can provide a large surface area for micro-organisms to colonise and can act as a trap organic material in the system, concentrating easily degradable material.

b) Prevent tramp oil accumulation, possibly by fitting of an oil skimming device. Tramp oil can form an oxygen impermeable layer on top of the coolant. This can result in oxygen depletion of the coolant promoting the growth of anaerobic bacteria, in particular sulfate reducing bacteria.

c) Ensure correct emulsion strength is used. This will ensure that the buffering capacity of the product is good and that an appropriate concentration of preservative is present in the product.

d) Maintenance of good pH control is key. An elevated pH will discourage the proliferation of micro-organisms, in particular the filamentous fungi.

e)   Do not allow the coolant to stagnate. Coolant stagnation promotes the growth of anaerobic bacteria.

f)   Regular monitoring and maintenance of the metalworking fluid can result in extended working life and provide an early warning of any potential problems. This early warning can result in saved time and money.

g)   Use products that are pre formulated with a bactericide and/ or a fungicide.

3   FACTORS AFFECTING THE SELECTION OF BIOCIDES FOR USE IN METALWORKING FLUIDS

Formulators have many factors to consider when selecting an appropriate preservation system for a candidate metalworking fluid. There are three broad areas to consider:

3.1   Biocidal Efficacy

One of the first issues to consider is whether it is desirable to have anti-bacterial and/or anti-fungal activity within the product. In practice, metalworking fluids are susceptible to attack by both bacteria and fungi. As a consequence, it is usually desirable to have both activities present in a formulation. There are a number of options for achieving this desired level of protection. Use a mixture of bactericides and fungicides to give a broad spectrum of anti microbial activity. Or, use a broad spectrum biocide that is efficacious against both bacteria and fungi.

It is also important to consider the dose of biocide required to give the desired longevity in-use and compare this to the cost of the preservative. In many metalworking fluids, the preservative system can be the single most expensive component in the formulation. Getting the balance between cost and efficacy is key.

3.2   Physicochemical Issues

Biocides are formulated into the concentrates of products at anywhere between 20–25 times the end-use concentration. The preservative system must be soluble in the concentrate which may be a mixture of oil and water. The concentrate may then be stored anywhere between three months to one year. During this period of time, the biocide must remain active if it is to be efficacious when the product is used. There are many components in a metalworking fluid formulation that can deactivate certain types of biocides. For example, the presence of primary amines in many products rapidly deactivate isothiazolinone biocides, making this class of biocides unsuitable for use in the vast majority of metalworking fluid concentrates.

Most metalworking fluids are buffered at about pH 9.0–9.5. Any biocide used must have long-term stability and be efficacious at these pH ranges. In addition to this, a degree of thermal stability is desirable. The main body of diluted metalworking fluid will maintain at a relatively constant temperature, usually about the ambient temperature. The product concentrate may be subject to significant temperature variation depending on how it is stored and the part of the world it is destined for.

It is highly recommended that both the efficacy and physicochemical compatibility of the preservation system in the candidate formulation are evaluated using recognised test protocols such as those published by the Institute of Petroleum and ASTM.

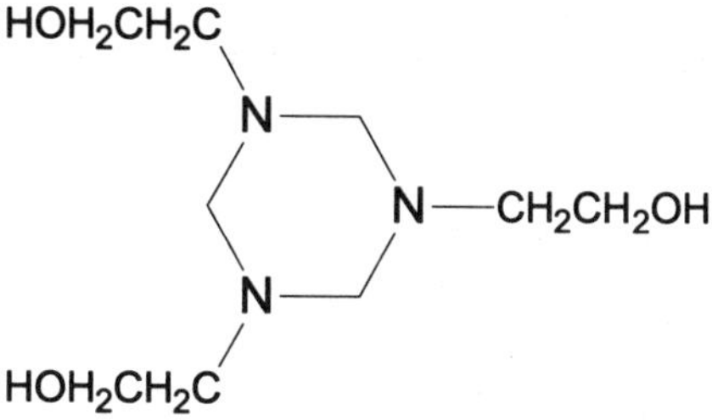

3.3 Non-Technical Factors

Increasingly, non-technical factors are playing a more important role in the selection of biocides for use in metalworking fluids. Within Europe, the use

Figure 2: Triazine

of many biocides is unrestricted, with national legislation varying significantly from country to country. However, some biocides are subject to labelling restrictions under the European Dangerous Substances Directive. This can impact significantly upon the use of biocides in some products. This is particularly the case in respect of labelling requirements in respect of potential skin sensitisers. Further, the advent of the European Biocidal Products Directive will almost certainly change the shape of the biocides market in Europe, having a direct impact on metalworking fluid formulators.

Customer preference is also an increasingly important consideration when selecting biocides for use in products. Large manufacturers such as Ford, Volvo, BMW and SKF, to name but a few, are producing lists of materials they do not want to see in their factories. Many biocides are on these lists. Currently, they are not prohibited, but this may only be a matter of time.

4 BIOCIDES USED IN METALWORKING FLUIDS

What follows is an examination of the typical biocidal chemistries used in metalworking fluids. The list is by no means exhaustive, but should give a flavour of what is available and the advantages and disadvantages of certain chemistries.

4.1 Formaldehyde Release Biocides.

Formaldehyde release biocides are perhaps the most commonly used biocides in metalworking fluids. One of the best know examples of this chemistry is hexahedron -1,3,5-tris(2-hydroxyethyl)-s-triazine, (see Figure 2).

This is a monoethanolamine formaldehyde condensate. It is cheap, compatible with most metalworking fluid concentrates and provides a valuable source of reserve alkalinity. It is bactericidal and is typically incorporated into metalworking fluid concentrates at between 2.5 to 3% by weight. Good antibacterial performance is seen at between 1200–1500 parts per million (ppm) in-use.

It does have a number of draw backs. It has poor thermal stability (a property common to most formaldehyde release biocides) and, in some instances, may cause blackening of metalworking fluid concentrates if heated above 50°C for a period of time. Recently, this active ingredient was placed on Annex 1 of the Dangerous Substances Directive having been identified as a potential skin sensitiser. This means that formulations containing efficacious levels of this class of triazine in them would have to be labelled with R43 – may cause sensitisation by skin contact. This is unacceptable to many UK customers. As this material is only bactericidal, it needs to be co-formulated with a fungicide to provide complete protection for a product.

Another class of formaldehyde release biocide are the oxizolidines. An example of a typical oxizolidine would be bis-(5,5'-dimethyl-1,3-oxazolidin-3-yl)-methane, (see Figure 3).

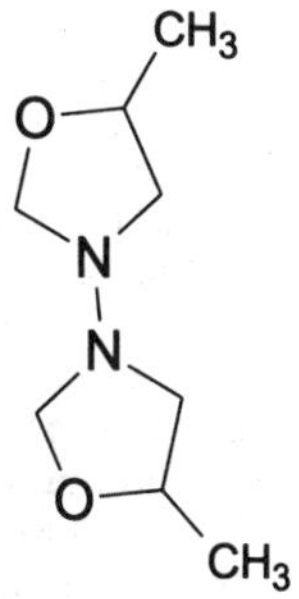

Figure 3: Oxizolidine

This material shares many of the advantages of the triazine biocides. They are relatively cheap, compatible with most formulations and provide a source of reserve alkalinity. In addition, at the time of writing, there are no requirements for labelling R43 with this class of material. It shares similar disadvantages to the triazines, possessing poor thermal stability. Oxizolidines also need to be co-formulated with a fungicide to provide complete protection for a product.

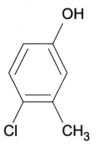

Figure 4: PCMC

4.2 Phenolic Biocides

Phenolic biocides are still used extensively throughout the metalworking industry to preserve aqueous based metalworking fluids. One of the most commonly used phenolic biocides is p-chloro-m-cresol, (see PCMC, Figure 4).

It is soluble in water to about 4g/l; it is more common to use this biocide as its sodium salt, improving its water solubility. It is incorporated into metalworking fluid concentrates at typically between 2 to 4% by weight.

It has a number of advantages, having a broad spectrum of activity against both bacteria and fungi. It does not require co-formulation with a separate fungicide. It exhibits good thermal stability and is free of formaldehyde.

Phenolics possess two major disadvantages for metal working fluid formulators. In our industry, there is an issue relating to phenolics in respect of their wastewater treat ability. Also, PCMC has been placed on Annex 1 of the Dangerous Substances Directive having been identified as a potential skin sensitiser. This means that formulations containing more than 1% by weight of PCMC in them would have to be labelled with R43 – may cause sensitisation by skin contact. This is unacceptable to many UK customers.

Figure 5: isothiazolinones (CMIT/MI)

4.3 Isothiazolinone Biocides

Another major chemistry often used are isothiazolinones. There are a number of different varieties of this chemistry, probably the most common form used is supplied as a mixture of 5-chloro-2-methyl-4-isothiazolin-3-one and 2-methyl-4-isothiazolin-3-one, (see Figure 5, usually in the ration of 3:1).

This chemistry has many advantages. Firstly, commercially available preparations are formaldehyde and phenol free. It is broad spectrum and rapid acting in terms of the biocides available to metalworking fluid users. They are incompatible with most metalworking fluid concentrates and cannot be formulated in to these. They are used principally for tankside addition. Usual treat rates are between 12 and 25 ppm active ingredient, being economic to use and efficacious at these concentrations. As they are unstable at the pH range commonly found in metalworking fluids, they are rapidly degraded and can disappear from a formulation in as little as forty-eight hours post addition.

Incompatibility with fluid formulations is one disadvantage of this chemistry, another is its potential to induce skin sensitisation. Benzisothiazolinone, (see Figure 6) is a chemically more stable isothiazolinone. This material can be incorporated into most metalworking fluid concentrates, over coming the stability problems seen with other isothiazolinones. However, this higher stability is reflected in higher "in-use" levels required to control micro-organisms. Typical "in-use" levels are between 200–400 ppm of active ingredient.

Figure 6: Benzisothiazolinone

4.4 Phenoxyalcohols

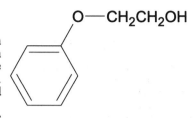

Figure 7: Phenoxyethanol

Phenoxyalcohols are a relatively new chemistry in metalworking fluids. They first appeared in specialised formulations back in the 1980's. They are a part of a newer generation of softer biocides. Phenoxyethanol is one of the most commonly found actives of this class of biocides, (see Figure 7).

They have many advantages, they are formaldehyde free, halogen free, demonstrate good thermal stability and have good skin compatibility. However, it demonstrates no anti fungal activity and requires co-formulation with a dedicated fungicide to give a broad spectrum of protection. Phenoxyethanol also has very high "in-use" levels, requiring between 3000ppm and 7000ppm some two to three times higher than more traditional biocides. A metalworking fluid concentrate formulated with this biocide may contain between 10 to 15% by weight of phenoxyethanol. This may result in an odour in-use which some may find offensive. As a consequence, phenoxyethanol preservation tends to be used in more specialised product types such as amine and boron free metalworking fluids.

5 FUNGICIDAL TECHNOLOGY

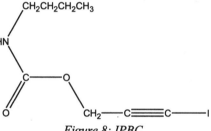

Figure 8: IPBC

Most of the chemistries described above demonstrate principally a bactericidal activity. Most metalworking fluids are susceptible to degradation by filamentous fungi. The majority of commercially available water extendible metalworking fluids will contain both a bactericide and a fungicide. There are two main fungicides used in metalworking fluids.

5.1 3-Iodopropargyl-N-butylcarbamate (IPBC)

IPBC, (see Figure 8) is used extensively by the wood preservation industry. It has an excellent toxicity profile and is extremely efficacious against fungi.

Currently, its use is restricted to the emulsifiable metalworking fluids. This material is sparingly soluble in water and cannot be used in the water soluble synthetic metalworking fluids. Good long-term anti fungal performance in metalworking fluids is seen at "in-use" levels between 100ppm and 200ppm of active ingredient. However, IPBC can be very difficult to formulate with, it is thermally unstable and can be unstable in the presence of some metalworking fluid components.

5.2 Sodium Pyrithione

The other major fungicide used in metalworking fluids is sodium pyrithione (Figure 9).

This fungicide is freely soluble in water and as a consequence can be used to provide fungicidal activity in water soluble synthetic metalworking fluids. It provides good protection at concentrations of 150ppm to 250ppm of active ingredient "in-use". Its main disadvantage is its tendency to chelate metals. Sodium pyrithione has a very high affinity for iron, leading to the discolouration of the metalworking fluid in the presence of iron. This can be overcome by co-formulation of this fungicide with a suitable sequestrant.

Figure 9: Sodium Pyrithione

6 CONCLUSION

The range of biocides suitable for use in metalworking fluids is limited to a small number of chemistries. Within these chemistries, there is a wide range of active ingredients available. Metalworking fluids vary enormously in their formulatory composition. As a consequence, any proposed anti-microbial package should be validated by challenge testing before a product is placed in a key customer account.

It is clear that the latter half of the 1990's has been characterised by two phenomena. Firstly, the regulatory pressure on active ingredients is increasing. Within Europe, the advent of the biocidal products directive could change the shape of the biocides market within Europe and potentially further afield. This piece of legislation is expected to reduce the number of commercially available active ingredients by at least 50%. Some estimates suggest an even greater reduction.

Secondly, customers are applying ever greater pressures on formulators to make less use of some chemistries, for example formaldehyde release biocides. Customers would prefer preservatives that are good for the environment and possess no adverse mammalian toxicity characteristics. This would require that new actives are developed. This is unlikely considering the new regulatory frameworks.

The existing range of actives are effective and provide good performance in metalworking fluids, newer chemistries however would be desirable.

AIRCRAFT TOILET SANITISING

T. F. Child, A. S. C. Whyle, B. Bell, T. Willems

Brent International PLC
Denbigh West, Bletchley
Milton Keynes MK1 1PB

1 INTRODUCTION

In 1944 Wellington bomber "R for Robert" was on a training mission in Scotland. During the course of the exercise the plane experienced difficulties and was forced to ditch into Loch Ness with the loss of one crewman. The bomber settled on the bottom of the lakebed and lay undisturbed for years. Divers discovered her 10 years ago in a remarkable state of preservation and a decision was taken to raise her from the lake and undertake a complete renovation of the aircraft. Among the items found on board was an "Elsan" toilet, undamaged and in excellent condition. The problems of maintaining and sanitising toilets on wartime aircraft were not key issues at that time, but today this aspect is extremely important in civil aviation.

"R for Robert" is shown in Figure 1 during renovation, note the bent props which suffered damage during the landing on water. The "Elsan" toilet is shown in Figure 2 with the trade markings clearly visible.

Figure 1 *"R for Robert" undergoing renovation*   **Figure 2** *The "Elsan"*

2 TYPES OF AIRCRAFT TOILET

There are two types of aircraft toilet used on board passenger-carrying aircraft today and each works on quite different principles.

2.1 Recirculating

Figure 3 shows what is going on beneath the flap valve of the toilet. It is a closed circulating system with main components being a toilet pan, a tank with pump, tube connections and an operating mechanism at the outside of the aircraft body to facilitate servicing. The mixture of water, urine, solids and sanitiser is filtered and pumped back to the toilet during each flush. The sanitiser contains disinfectants, perfume for odour masking and dark blue dye to improve appearance of the flushing liquid. Emptying and rinsing of the toilet tank are done from the ground. The trucks used to empty the tanks and transport the contents away are affectionately known as "honey wagons".

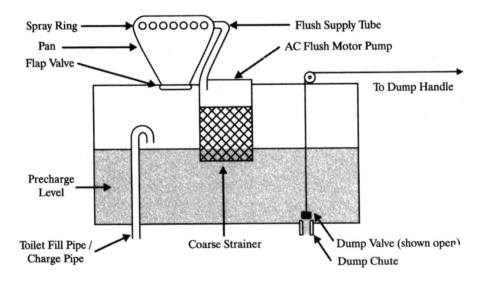

Figure 3 *Re-circulating toilet schematic*

Initially, the tank is charged with 25% of its volume with water and the appropriate amount of sanitiser. Products are introduced by adding a water-soluble sachet containing liquid or powder sanitisers or by adding a liquid product directly from a plastic bottle pre-filled with the correct amount. This procedure accurately controls dosage rates and limits worker exposure. Products are diluted more than 400 times and are required to kill all pathogens within 15 minutes.

The disadvantages of this type of toilet system are the high usage of sanitiser and high volume of liquids/solids for subsequent disposal. This led to the development of the vacuum toilet system.

2.2 Vacuum

All toilets on an aircraft are connected with one or more central tanks through vacuum pipes. This is illustrated in Figure 4. Solids are transported to the tanks by high vacuum, creating an air displacement of over 60 m per second. Up to flight altitudes of 4-5000 m, an electrical pump provides the necessary low pressure. In altitudes exceeding this level, the low pressure outside the aircraft provides the necessary pressure difference amounting to 0.5-0.6 bar.

A central vacuum-system controller regulates opening and closing of the drain valves and the supply of water. Due to this controlled system, one rinse of the toilet needs only 0.25-0.30 litre of water (a normal domestic toilet needs 7-9 litre).

Solids and liquids are collected in special lightweight tanks. The number and size of the tanks will depend on the size of the aircraft, typically 220 litre for an Airbus A320/321 and 700-1100 litre for an Airbus A330/340 (2-3 tanks). Similar to the re-circulation toilet, tanks are charged with 25% of their total volume with the requisite amount of water and sanitiser.

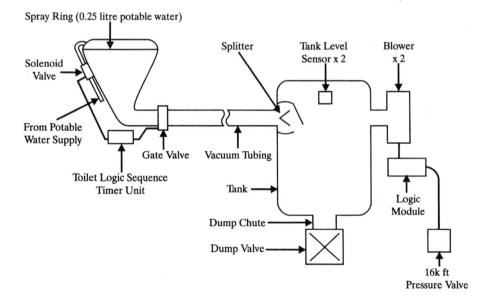

Figure 4 *Vacuum toilet schematic*

A central disposal pipe in the service panel at the bottom of the aircraft body enables the quick and clean discharge of the toilet tanks. Special ball valves avoid building of "blue ice" during flight or on the ground. After disposal, the system is cleaned through spray nozzles and the cleaning truck discharges the effluent directly into the public sewage system.

3 MATERIALS' TESTING

Aircraft toilet sanitisers may come into contact with a variety of different materials through spillage and leakage and therefore materials' testing is essential on neat and diluted sanitiser solutions. Problems have occurred in the past where electrical connectors have been short-circuited by leakage in the toilet area and caused loss of aircraft trim during flight. A typical list of materials tested is shown below:

- Stainless steel, aluminium alloys
- Composite, titanium
- Sealants, adhesives
- Plastics, cabling

4 ESSENTIAL PRODUCT PERFORMANCE REQUIREMENTS

Products are required to provide a total kill of bacteria and viruses during 15 minutes contact with the sanitiser. This challenge might seem relatively straightforward, however, when you take into account that this performance requirement is needed up to 24 hours (the maximum time between discharge), the formulators' options become quickly exhausted.

Combined with the above requirements, the product should control/eliminate scale formation, contain surfactants to clean the system and be low-foaming. Toilet sanitisers are required to control odour whilst being of relatively low odour themselves. Although rapid bacteria kill is essential, the discharged product must be treatable in a domestic sewage system without interfering with its efficiency.

5 PRODUCT TESTING PROTOCOLS

A further complication is that different airlines demand different test protocol. Some use raw sewage in tests, some do not. Others use a combination of tests to simulate the different conditions experienced in toilet tanks. British Airways for example use a simple challenge test that is useful for screening.

5.1 British Airways Challenge Test

UHT milk is used to simulate the "organic loading". Using a 1:200 dilution of the neat product, test solutions are prepared to simulate "initial" charge at 1:400 and "final" charge at 1:1600; equating to an aircraft toilet tank when completely full. Test suspensions are set up using 250ml sterile glass sampling bottles. The suspensions are incubated at 24 °C and 1 cm^3 aliquots removed at 5, 10, 15 and 20 minutes.

Duplicate counts are made using serial dilutions up to 10^{-6} and "drop" plates. Solutions are then spotted onto blood agar plates and incubated at 37 °C for 18 hours after which the number of colony forming units is determined. To pass the "BA Challenge Test" there must be no growth from the aliquots taken at 15 minutes or more from the 1:400 and 1:1600 dilutions.

Whereas this test is considered by some to be inadequate for an accurate indication of sanitiser effectiveness, we have found it useful when screening a large number of formulations.

5.2 Other Test Protocols

It is not intended to go into details about the various test protocols from different airlines. Suffice it to say that they differ considerably; for example, Air Canada uses raw sewage in their test method whereas Lufthansa uses raw sewage with additional bacteria and fungi. It is argued that these methods give a more realistic indication of sanitiser effectiveness, which is probably true, although these tests are preferably undertaken by a specialist organisation (e.g., Hygiene Institute Hamburg).

6 CHEMISTRY OF TOILET SANITISERS

Increasing environmental pressure and worker safety considerations are contributing to changes in the industry. The old systems of aldehydes and phenols are no longer acceptable. Up until quite recently toilet sanitisers were formulated with glutaraldehyde, formaldehyde or formaldehyde donors. Glutaraldehyde has been a favourite with the formulators for some time because of its formulation flexibility, although mixtures of amines have also found favour because of synergistic effects and virucidal activity.

Quaternary-based materials have been used as alternatives to aldehydes and although they provide acceptable kill rates and odour control, the subsequent disposal of effluent presents heightened challenges. The growing concern over virus survival in toilet tanks has also caused some rethinking of the technology and alternative active materials and their combinations are constantly being researched.

7 THE IDEAL PRODUCT

The ideal product may not exist with today's technology, but an attempt has been made here to list the most desired properties of an aircraft toilet sanitiser to satisfy customers' needs and at the same time maintain safety for workers and the environment:

- Meets airline specifications and all legislative requirements
- Rapid and long acting
- Effective against gram negative bacteria, fungi and viruses
- Compatible with metal, sealants, cabling, plastic and composites
- No chlorides, volatile ingredients, heavy metals, phenols
- Biodegradable
- Treatable in domestic sewage systems
- No aldehydes or quaternary ammonium compounds
- Compatible with anionics
- Low risks on worker contact
- Non-flammable
- Easy, accurate and safe dosage
- Long-lasting odour control
- Control of scale build-up in re-circulating toilet lines
- Non-staining
- Long-term stability

8 THE FUTURE

According to a study by the World Health Organisation, viruses have been found to survive chemical treatment in aircraft toilets. Researches reported that up to half the samples of waste pumped from on-board toilets contained viruses despite the use of toilet sanitiser. All the viruses isolated from the waste were associated with enteric disorders. This indicates that other viruses such as those that cause hepatitis could also survive. As a consequence, future products will need to be even more effective as it is believed that between 1-10% of viruses survive to be potentially discharged into the environment.

On the other hand, we need to ensure that stronger sanitisers do not damage the health of workers during discharge of toilet tanks nor upset normal biodegradation processes in effluent treatment plants.

Swimming Pools

THE USE OF ALGICIDES IN SWIMMING POOLS

Yves L. Verlinden

European Formulator Marketing Coordinator, Buckman Laboratories, SA Belgium

1 INTRODUCTION

What we wrongly learned in school, water is not just H_2O, two atoms of hydrogen combined with one atom of oxygen. If we look at water under the microscope, we see that water is a liquid with a lot of microbiological growth, bugs as we say. In general, these bugs are divided into three groups:

- Bacteria
- Fungi
- Algae

It is the last group we will expand in detail.

Algae can be useful and harmful at the same time.

Useful:

- As food (mainly sea-algae) in Japan, China, Hawaii and Philippines.
- They are used for the production of Agar-agar
- And they are used in the pharmaceutical – and cosmetics industries.

Harmful:

- Annoying and troublesome in swimming pools
- Blocking filters
- Can be toxic (e.g. mussel poisoning).

That is why in cooling water towers and in swimming pools algicides are used.

2 MORPHOLOGY

Typical for algae is that they contain a pigment called Chlorophyll which is responsible for photosynthesis. We consequently can decide that algae belong to the family of plants.

Algae need three basic elements so that they can develop and multiply: air, water and light. The photosynthesis works as follows: the cell material is built by CO_2 (in water present as bicarbonates HCO_3^-) as carbon donor, the water itself as electron donor and light as energy source, whereby the chlorophyll acts as a catalyst. The chemical substances in the water, like Ca^{2+}, Mg^{2+}, silicium, Fe^{2+}, Mn^{2+}, phosphates, nitrates, etc will influence the structure and growth of the algae. Algae do not need organic material for their growth and multiplication, although phosphates encourage the algal growth. Algae can appear as 'unicellular' structures with small dimensions like bacteria, or they can multiply until they are long filamentous threads or fibres called 'seaweed' like we find on rocks, piers and bulwarks. Algae can be divided into the following groups:

- Chlorophytes (or green weeds), such as *chlorella, Chlorococcum* and *Ulothrix.*

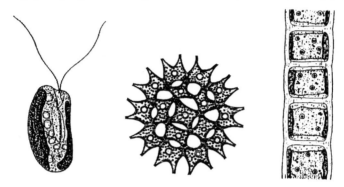

- Brown Algae, such as *rhodophyta* and *phaeophyta*. They consist of numerous compilations of thread-like cells, which contain specific brown coloured pigments, like Fe^{2+} and Mn^{2+}.

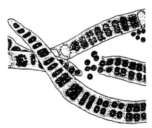

- Blue Algae, such as *cyanophyta,* appear as unicellular threads; they have a typical blue colour and they are able to bind elementary nitrogen.

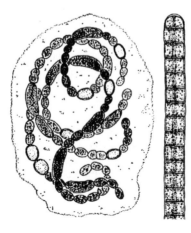

- Lichen is in fact a symbiose between a weed and a fungus. They are not found in swimming pools. Mostly they are present in open recirculating cooling water systems.

2.1 Where do the Algae in the Swimming Pools come from?

A large population of algae is present on the surface of and within most soils. When the soil dries it is swept into the atmosphere by winds, and distributed far and wide as dust. When dropped in the water of a pool, they start multiplying by using photosynthesis, explained above. Many algae produce within their cells *malodorous oils* and *cyclic alcohols,* that are released when the organism dies, causing odour problems. Using a proper microbicidal treatment, the algae is killed, dry up and disappear completely.

3   DISINFECTION: MICROBICIDAL TREATMENT FOR POOLS

Strictly, disinfection means removing the risk of infection, but in the context of swimming pools the water cannot be sterile all the time. A few living, but normally harmless microorganisms will always be present, especially in open pools. Disinfection aims to keep their number to an acceptable minimum and to ensure that any harmful organisms entering the pool water are rapidly inactivated and eliminated so that the water will not transmit infection to the bathers.

Proper disinfection is achieved by maintaining the correct concentration of disinfectant in the poolwater. Also FILTRATION is extremely important.

For disinfection the water should be clear and free of suspended material because they will react with some disinfectants. This will be discussed under quaternary ammonium salts.

The disinfectant must be given TIME to kill the microbes.

3.1 Choosing a Disinfectant

When disinfection is discussed in different magazines and manuals, people always speak about mainstream disinfectants like chlorination or bromination, oxidation and ozonation. But much less, if at all, about algicides and non-oxidizing biocides. Therefore this discussion will be more on the use of algicides and will prove that the use of chlorine is not always that effective. In terms of effectiveness in use there is little to choose between chlorine gas or the hypochlorite form with regard to microorganisms (bacterial or algae) in swimming pools. Chlorine gas is a very effective but dangerous bactericide when not used correctly. Also sodium and calcium hypochlorites are able to produce dangerous levels of chlorine gas if accidently mixed with acids. Chlorine is less effective on algae. Chlorine gas is clearly distinct from sodium and calcium hypochlorites, but their action in pool water treatment is similar. Chlorine is a powerful oxidizing agent and much of the polluting matter introduced to the pool water by bathers is capable of oxidation. Most of the loss of chlorine is to such oxidation. This loss must continually be made good to ensure that there is a residual level of chlorine in the pool, to deal with fresh pollution. Chlorine gas dissolves in water and reacts reversibly with it:

$$Cl_2 + H_2O \leftrightarrow HOCl + HCl \tag{1}$$

The hypochlorous acid (HOCl) is the active agent responsible for the disinfection. The hypochlorites also produce hypochlorous acid in water;

$$NaOCl + H_2O \leftrightarrow HOCl + Na^+ + OH^- \tag{2}$$

$$Ca(OCl)_2 + 2H_2O \leftrightarrow HOCl + Ca^{2+} + OH^- \tag{3}$$

Hypochlorous acid is a weak acid and reacts further in water:

$$HOCl \leftrightarrow H+ + OCl^- \tag{4}$$

The sum of hypochlorous acid and hypochlorine ion is what is measured as free residual chlorine. The hypochlorous acid is a 20 times stronger biocide that the OCl-form.

4   DISADVANTAGES OF CHLORINE GAS/HYPOCHLORITES

1. pH of Pool Water
From equation (4) above, we can see that, if pH is decreasing (meaning more H^+ is present), the equilibrium is pushed to the left; so hypochlorous acid, the active agent, also increases and hypochlorites are decreasing.   And hypochlorous acid is the strongest disinfectant of the two.   Thus, the pH of the water has an effect on the disinfectant grade.

When pool water is treated according to the rules (pH between 6.8 and 7.2) we can deduct from the graphic hereunder that the disinfection grade or '% value of hypochloric acid' is going down: we are losing efficacy!

pH effect on disinfection

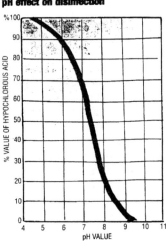

Here we see, at normal pool pH of 6.8-7.2 we lose about 40-50% of hypochlorous acid, responsible for the disinfection.

2. Formation of dangerous chloramines
Chlorine gas and hypochlorites react with pollutants in the water to form other products. As a result their ability to kill microorganisms is strongly reduced.

A major pollutant is ammonia (NH_3), which is continually added to pool water through the decomposition of the urea in nitrogenous products (urine, sweat etc) introduced by bathers.

The products of this reaction are chloramines, which are measured as 'combined chlorine'.

The reaction between chlorine and ammonia consists as three stages:

- First the formation of monochloramines:

$$Cl_2 + NH_3 \rightarrow HCl + NH_2Cl \tag{1}$$

- The second stage is the formation of dichloramines: the monochloramines react further with chlorine. Dichloramine irritates the eyes and nose.

$$NH_2Cl + Cl_2 \rightarrow HCl + NHCl_2 \tag{2}$$

- And finally, trichloramines are formed when chlorine reacts further with the dichloramines.

$$NHCl_2 + Cl_2 \rightarrow HCl + NCl_3 \tag{3}$$

Trichloramine is the most irritant of the chloramines; together with dichloramine it is largely responsible for the chlrorine odours and eye irritation.

This reaction is called the 'breakpoint chlorination': the combined chlorine level, which was rising as more chlorine was added, now drops suddenly. Free chlorine then rises without a corresponding rise in combined chlorine. This indicates that the pool pollution has been successfully oxidized by chlorine.

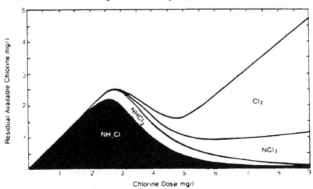

3. *The presence of iron (Fe^{2+}) and manganese (Mn^{2+}) in pool water*
One will loose biocidal activity due to the presence of iron and manganese in the pool water because chlorine and chlorine related products will oxidize these cations:

$$Fe^{2+} + Cl_2 \rightarrow Fe^{3+} + Cl^-$$

4. *Possibility of corrosion of stainless steel parts in pool*
By chlorination, we observe that we are producing chlorides. We see that in the hypochlorous reaction, as well as in the nitrogenous reaction, where HCl (chloric acid) is formed.

Chlorides are responsible for the pitting corrosion of steel parts. Normal carbon steel can stand 1000 ppm of chlorides ($=1000$ g M^{-3}), but stainless steel starts to corrode severely from 100 ppm on! Attention for ladders, illumination sets etc.

5. *Attack of pool particles by chlorine*
Indeed, it has been proven that chlorine attacks the cement mortar joints between the tiles of a swimming pool.

So we see the biggest disadvantage by using chlorine (and chlorine releasing products) and hypochlorites, is the minor efficiency at higher pH level and in presence of polluting nitrogenous products, and the fact is it is poisonous.

Therefore some quats, used as an algicide, can reduce the use of chlorine and even give a synergism and increased efficacy.

5 THE USE OF ALGICIDES IN POOLS

Outdoor pools and some indoor pools, exposed to sunlight, may experience problems with algal growth, particularly where the pool hydraulics are poor. Even phosphates can encourage algal growth. Chlorine is an effective algicide but will not compensate for the effect of poor hydraulics. Growth can develop very quickly given warmth and sunlight. Even just after a thunderstorm we see that algae bloom.

The longer it persists the more difficult it is to remove. So it is important to act immediately there are signs of algal growth. That first sign is most probably the greenish colour appearing in the pool water.

It will be necessary to use a decent algicide. Most of the approved algicides are based on quaternary ammonium salts or polyoxyimino compounds. Sometimes copper based, when black algae are present.

The main treatment is getting the pool management right from the start. Algal growth should not happen if the pool is operated properly.

Available algicides on the market:
- Quaternary ammonium salts
  - Normal quats or benzalkoniumchlorides
  - Polyquats (polymerized quaternary ammonium salts)
- Silver/colloidal silver
- Copper sulfate
- PHMB

The overall advantage of quaternary compounds is that their dosages are very low: 1-2 ppm per week only, where we see that Baquacil, although a good bactericide, must be dosed in such a way that we obtain about 100 ppm to kill the algae and some 40 ppm of hydrogen peroxide (H_2O_2).

Another advantage of quats and polyquats is that they have a low toxicity. We will discuss Oral LD_{50} data later.

5.1 Common Algicides

5.1.1 The family of Benzalkonium chlorides. The general formula is:

$$CH_3$$
$$\backslash \ Cl^-$$
$$N^+-CH_2-\langle \bigcirc \rangle$$
$$/ \backslash$$
$$CH_3 \ \ R$$

Quaternary ammonium salts are organically substituted nitrogen compounds in which the nitrogen atom is pentavalent. Four of the substituents are alkyl, aryl or heterocyclic radicals and the fifth is an anion (a cationic charge). This anion is mostly chloride. Therefore we call them benzalkonium chlorides.

To be active as a microbicide at least one of the organic radicals must have 8 to 18 carbons in the chain.

Quaternary compounds attack the phospholipid material in the cytoplasmic membrane of the microorganisms, causing 'lysis' (= deterioration) of the cytoplasm and death of the cell. Cationic chemicals are moreover 'toxic' because they form electrostatic bonds with negatively charged sites in the cell. The absorption of cationic surfactants disorganizes the semi-permeable membrane of the cell, causing cytological damage and leakages.

Calcium and magnesium reduce the toxicity of quaternary compounds by competing with them for acidic sites.

The benzalkonium chlorides are very foaming and can cause severe foam problems in spas and pools.

5.1.2 The Polyquats or polymerized quaternary salt. If we study the list of algicides approved by the DETR Committee on pages 107 and 108 in the book 'Swimming Pool Water' published by the Pool water treatment Advisory group, we observe the Polyquats WSCP (or BUSAN 77), WSCP-2 (or BUSAN 79) and APCA (or BUSAN 1055) and their impact on 'Hazard' and 'Incompatibility data'. Indeed, these quats have a very low toxicity and are compatible with most swimming pool chemicals used.

The company Buckman Laboratories came out for the first time some 25 years ago with Polyquats. They were obtained by a reaction with a tertiary amine TMEDA (tetramethylethylene diamine), where the methyl groups can react easily with an ether.

Their molecules and structures follow. (WSCP = Water Soluble Cationic Polymer).

WSCP:

$$\left[-O - (CH_2)_2 - \underset{\underset{CH_3}{|}}{\overset{\overset{CH_3 \quad Cl^-}{|}}{N^+}} - (CH_2)_2 - \underset{\underset{CH_3}{|}}{\overset{\overset{CH_3 \quad Cl^-}{|}}{N^+}} - (CH_2)_2 - \right]_n$$

WSCP-2:

$$\left[-\underset{\underset{}{\overset{\overset{OH}{|}}{CH}}}{} - CH_2 - \underset{\underset{CH_3}{|}}{\overset{\overset{CH_3 \quad Cl^-}{|}}{N^+}} - (CH_2)_2 - \underset{\underset{CH_3}{|}}{\overset{\overset{CH_3 \quad Cl^-}{|}}{N^+}} - CH_2 - \right]_n$$

We see that only two carbon atoms are between the nitrogen atoms: the lower the number of carbon atoms in these chains, the more algicidal effect and the more water soluble they are.

APCA:

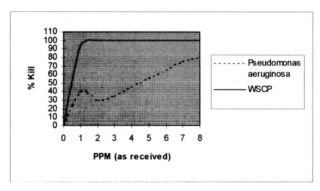

Differences between normal quats and polyquats.

(a) Polyquats, with a low number of carbon atoms between the nitrogen are absolutely non foaming.   The number of carbon atoms per nitrogen atom is influencing the tensioactivity of the product. The lower this figure is, the lower the foaming effect has on it. E.g. for a dimethyl alky (C18) benzalkonium chloride we have 18+1+1+6 = 26 carbon atoms per nitrogen. For WSCP and WSCP-2, we have 2+1+1 = 4 carbon atoms per nitrogen.

(b) Polyquats have the advantage, due to their extremely high cationic charge, that they are excellent flocculants. They are flocculating organic substances like sun tan oil, body lotion and dead algae. Advantage: when 'properly' using WSCP or WSCP-2 (once a week 1-2 ppm) there is no need for extra flocculant.

(c) Polyquats (like WSCP and WSCP-2) when acting as a flocculant, have the intention of forming big porous flocs, so that the filters will not be blocked easily. Moreover, these quats are extremely efficient against 'Pathogenic Bacteria' like Shigella, Pseudomonas and Klebsiella, which can develop on badly maintained backwash filters. WSCP e.g. show a 99.99% kill at only 1.5-2 ppm in demineralized water and 5 ppm in artificial pond water. Benzalkonium chlorides are also effective on these bacteria, but from the graphic below, we learn that at a dosage of 1.5 ppm they are at their maximum activity (only 65% kill) and at dosages of 8 ppm, they still are at 75% kill. The graph represents percentage of organism kill versus ppm of biocide used. The full line is WSCP.

(d) Polyquats have a lower toxicity than normal quats. Still all quats have a low toxicity grade:

The Oral LD_{50} data:

Normal quat: 234 mg kg^{-1}
Polyquat: 3000 mg kg^{-1}
Aspirin: 1240 mg kg^{-1}

That means that polyquats are some 2-3 times less toxic than aspirin.

(e) Polyquats are chlorine stable, up to 1 ppm of free chlorine in the pool. And this is a lot. In the German Public pools, the chlorine content is maintained between 0.3 and 0.6 ppm free chlorine.

(f) Normal quats or benzalkonium chlorides are better disinfectants. This means they are active after 30 seconds where the polyquats need a contact time of 4-6 hours. Of course, this has no negative effect on the efficacy of the polyquats, because they are added on a constant base (once per week) and they remain active over a much longer time.

(g) Polyquats showed a synergism with copper ions when used against the very severe 'black algae'. A bloom of cyanobacteria was found in an open storage reservoir. The bloom which had over 100,000 cells mL^{-1} of Schizothrix, Plectonema, Phormidium and Lyngbya was dosed with copper sulfate. In another case a bloom of Microcystis was treated. In both cases, after three weeks of its appearance, the water supply authority dosed the reservoir by aerally spreading 1 ppm of copper sulfate and 2 ppm of polyquat, which killed the blooms.

Attention must be paid to the fact that copper sulfate can cause considerable black-green staining on certain liners and even on tiles and marble.

Polyquats can easily be mixed with copper sulfate. The rule is first dilute the quat and subsequently add the copper sulfate solution afterwards.

(h) Polyquats, like WSCP and WSCP-2, showed an interesting synergism with oxidants like chlorine, chlorine dioxide and peroxides.

5.1.3 Silver/Colloidal Silver (Ag$^+$). Silver has a slow inhibiting effect on bacterial and algicidal growth. It is a very expensive treatment and certainly not cost effective. If the silver or colloidal silver is not stabilized, it is not compatible with quats and polyquats. It will react with the chlorides of the quats.

5.1.4 Copper Sulfate (CuSO$_4$). Copper sulfate as already discussed under 5.1.2(g), is an excellent algicide, but can give considerable black-green staining on some liners, tiles and marble. For safety and health reasons, in swimming pools, it should not exceed 1 mg l^{-1} (or 1 g M^{-3}). There are claims that ions and halogens, ions and quats together disinfect more effectively than the sum of their disinfectant powers separately. That is called synergism.

5.1.5 PHMB (or PolyHexaMethyleneBiguanide). Molecule 'invented' by ICI and was registered and promoted under the name Baquacil.

PHMB is very toxic to fish and aquatic life. It is moreover irritating to skin and may cause sensitization by skin contact. It can cause irritation to the eyes, nose and respiratory tract. The PHMB is not compatible with most common swimming pool chemicals. Not compatible with chlorine and chlorinated chemicals and bromine donors. Not compatible with ionic sterilizers, copper based QAC-algicides, anionic detergents, water softening chemicals, persulfate oxidants etc. The defence of the 'inventors' of PHMB is that one should not combine it with other biocides because it should be a bactericide/algicide. But the algicidal properties of PHMB are very weak: in brochures and manuals the dose is 200 ppm.

WSCP Enhances the Activity of Oxidizing Microbicides in Swimming Pools

New laboratory data has proven that less chlorine is required to kill bacteria when WSCP is present in a swimming pool. The graph below shows the dramatic reduction in MIC* for three common oxidizers—hydrogen peroxide, calcium hypochlorite, and chlorine dioxide—that can be achieved when WSCP is used.

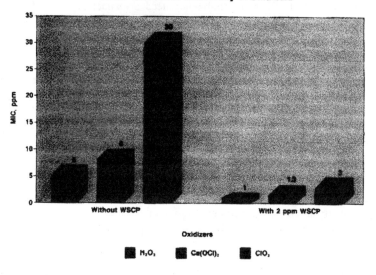

WSCP Enhances the Activity of Oxidizers

Results based on Pseudomonas at pH 8.0

The presence of WSCP at only 2 ppm can reduce by 80% or more the chlorine needed to kill bacteria. Just one more reason why WSCP is the algicide of choice in swimming pools.

*MIC (minimum inhibitory concentration) is a common laboratory measure of the effectiveness of a biocide. A lower MIC number means it requires less biocide to kill the bacteria used in the test.

It is important to note that French swimming pool formulators found empirically that PHMB is compatible with some polyquats with short chains between the nitrogen atoms, like WSCP, WSCP-2 and APCA. The combination polyquat/PHMB has now been on the market in France for many years and proved its quality and appearance.

5.1.6 Simazine. Simazine is a triazine-herbicide (for crops, ornamental plants etc.). Old information (1983) said that Simazine could be used as an algicide in swimming pools. Indeed it has some interesting algicidal properties. But according to the EPA, all triazines are classified as possible human carcinogens (Group C) based on an increase in mammary glands tumours in female laboratory animals.

Note: According to several studies performed in the USA, it seems that 'benzalkonium chloride' where the alkyl group R contains only 12 C-atoms (e.g. dodecyl benzylammonium chloride) is effective against the growth of Cryptosporidium spp.

THE FUTURE PROSPECTS FOR CHEMICAL BIOCIDES FOR POOL WATER TREATMENT

Simon J Judd

School of Water Sciences, Cranfield University, Beds. MK43 0AL, UK

ABSTRACT

The use of chemical biocides is ubiquitous in swimming pool water treatment. The choice of exact chemical employed may be guided by legislation (or guidelines) based on water quality, bather comfort and cost. The first two of these are ultimately related, since legislation is usually promulgated to preserve bather comfort and health. Of key interest, therefore, is the propensity for disinfection byproduct (DBP) formation within the swimming pool environment, since these form the basis of water quality legislation due to their recognised health effects.

Trends in chemical treatment of swimming pool water across Europe are outlined. These relate entirely to the use of "dry" chlorine, comprising stabilised chlorine and calcium hypochlorite. Results from a recent UK survey are also summarised.

The results of recent trials conducted at the Cranfield University pilot-scale swimming pool are also presented. This research is based largely on chlorination using hypochlorite, but includes other technologies. Results give a clear indication of the extent to which chlorinated byproducts within the swimming pool, as well as the relative levels of DBPs formed between bromination and chlorination. The implications of this research on the future of the use of chemical biocides in swimming pools is considered.

1 TRENDS IN DISINFECTANT USAGE

1.1 Worldwide/Europe, dry chlorine

"Dry chlorine" is the term given to chlorine-based disinfectant chemicals provided as powders, and comprise mainly calcium hypochlorite ($Ca(OCl)_2$) and the chlorinated isocyanurates. These two products, together with sodium hypochlorite, have been estimated to account for 90% of pool chemicals. Dry chlorine is used extensively in private pools, whilst sodium hypochlorite is extensively used, at least in Europe, in public and municipal pools. However, municipal pools in, for instance, Italy, France and Spain all use the chlorinated isocyanurates. The dry chlorine products (Table 1) have the advantage of ease of handling and dosing, as well as contributing high available chlorine levels in concentrated form.

Table 1: *Pool disinfectant quantities, dry chlorine*

	Ca(OCl)$_2$		Chlorinated isocyanurates	
	Market, kT	*% used in pools*	*Market, kT*	*% used in pools*
Worldwide	260	75	130-145	67
Europe	6-8	75	20-25	85

Growth through the 1990s for calcium hypochlorite was at around 3.5 percent per year with, globally, some 75 % of Ca(OCl)$_2$ is used in residential swimming pool sanitation and 25 % in municipal and industrial bleaching and sanitation. The use of calcium hypochlorite as a shock treatment in swimming pools as well as in specialised chemical feeding systems is increasing with pool construction. Typically it is used from "mid-season" in sunny climates when the isocyanuric acid level in private pools is approaching its limit and it is thus inappropriate to add chlorinated isocyanurates. However, the use of Ca(OCl)$_2$ promotes calcium scaling, a clearly undesirable – although readily treatable – side effect.

Chlorinated isocyanurates are used as sanitising chemicals and as bleaching agents. Trichloroisocyanuric acid (TCCA) is the principle product of the family and has typically 90% available chlorine content. Its slow solubility in water gives it the advantage of longer-term effectiveness in pool use. It is also known as trichloro (1,3,5) triazine-s-trione. The European Chemical Bureau refers to it as *symclosene*. In the industry, most producers refer to it as *trichlor*. The sodium salts of TCCA are more soluble and less acidic, but have a lower available chlorine content. Sodium dichloroisocyanurate (SDDC) exists in hydrated and di-hydrated product forms according to the amount of drying carried out on the wet slurry during processing. The di-hydrated form reduces the fire hazard associated with the mono-hydrate but also the amount of available chlorine. Its rapid aqueous solubility enables use in cooler pools or for "shock" treatment, but in particular lends itself to detergency use. It is also known as (1,3,5) triazine-trione, 1,3-dichloro, sodium salt. The European Chemical Bureau refers to it as *troclosene sodium*. In the industry, most producers refer to it as *dichlor*. Some 60% of global demand for chlorinated isocyanurates is for TCCA and 40% is for SDDC.

TCCA hydrolyses in water to form hypochlorous acid and cyanuric acid. The solubility is relatively slow but increases with increasing temperature. The dissociated cyanuric acid acts as a "sunscreen" for the chlorine, effectively maintaining available chlorine levels even under the impact of the sun (UV light and heat). This is why it is TCCA that is mainly used as a source of chlorine in swimming pool treatment. In particular, the use is in private swimming pools, where there is a relatively small ingress of bacterial bodies, and that normally in periods of warm weather. In contrast, the use of TCCA or SDCC in public (especially indoor) pools is strictly limited and even, in Germany and in some US states, forbidden. Although German public pool regulations exclude the use of chlorinated isocyanurates and of brominated compounds gaseous chlorine is stilled used in some municipal pools, although there is an increasing safety debate about this. It seems likely that Germany will follow other European countries in adopting sodium hypochlorite use, as this has demonstrably safely used over the years.

An early start to the pool season stimulates demand, enabling drawing down of stocks and the potential for pricing improvement. Correspondingly a slow start can put a strain on pricing. A detailed volume breakdown on an annual basis is difficult to determine. Other factors that influence markets (Table 2) include pool size. On balance it seems difficult to believe that more private pools exist in Germany than in France given the weighting of French pools to southern regions. On the other hand, this may reflect on both the smaller size of the German pools (30-50m^3 versus over 100m^3 for a French pool and 70-140m^3 in a Spanish pool) and the lack of any effective German coastline (apart from the Baltic, which is largely unswimmable). The 600,000 pools quoted by the pool association BSSW may be exaggerated by the inclusion of above ground and "splash" pools, and a more reasonable number is probably 230,000 pools. With a higher expectation of sunshine, Spanish consumption of chlorinated isocyanurates is the largest in Europe, ahead of France and Italy and well ahead of Germany. In round terms, growth for chlorinated isocyanurates, which clearly depends upon the global swimming pool population, has been occurring at a mean of 2.5-3.5% per annum, but for the seasonal reasons described the growth is not linear.

Table 2: *Pools (number, 000's) and growth*

Country	Public	Private	Total	Source (date)	Growth, %pa
UK	40	115	155	SPATA (1998)	4
Spain	-	-	350	ATEP (1995)	5
Germany	6.6	600[1]	607	BSSW (1998)	3
Belgium	0.8	50	50.8[2]	BSKZ	2
France	-	-	460	FNCESEL (1998)	4

[1]possibly nearer 230k, if "splash" pool excluded: 600k outdoor pools of which 310k sunk in ground, 290k "other" of which 167k outdoor and 123k indoor. Public pools around 6,6k of which 1820 schools, 1950 indoor, 2900 outdoor.
[2]estimate only by BSKZ President – no official statistics available. All 800 public pools use NaOCl.

A survey was recently conducted by Cranfield University School of Water Sciences in co-operation with the Pool Water Treatment Advisory Group to identify current trends in UK pool water treatment practice. About 500 pools, largely publically-run, were surveyed. A similar survey was conducted in 1989, and results (Table 3) indicate a shift towards both calcium hypochlorite and ozonation. In almost all cases other than chlorinated isocyanurates, where a

Table 3: *% UK disinfectant use*

Disinfectant	1989	1997
Sodium hypochlorite	55	37
Calcium hypochlorite	19	32
Chlorinated isocyanurates	?	3
Ozone plus hypochlorite	6	14
Bromination	1	2
Ultraviolet irradiation	0	2
Other	<19	10

higher measured chlorine residual is required to account for the proportion associated with the isocyanuric acid, the reported chlorine residual is kept below 2.5 ppm (Figure 1); three-quarters of ozonated pools and all UV-treated pools reported an operating chlorine residual of less than 1 ppm.

The main objective of the survey was to ascertain attitudes and perception of pool operators towards pool water treatment. To this end, the survey included two key questions:

Please indicate which three of the following six issues are the most important when choosing a disinfectant. (Bather comfort, cost, environmental issues, safety, suppliers service, water quality and clarity).

Please indicate which six of the following issues are the most important when considering the comfort and well-being of users and staff. (Air temperature, bacteria, chloramine levels in the water, chlorinous smells, clear water, fresh air, humidity, pH, residual disinfectant level, trihalomethanes, water temperature, water balance)

Results (Figures 2 and 3) indicate that choice of disinfectant appears to be largely dictated by safety and comfort of bathers, and also their perception of the pool water (i.e. its clarity and colour), with cost being of rather less concern (Figure 2). The responses to the second question were more varied (Figure 3), but tended to emphasise water quality aspects of the pool environment (Figure 4). In fact, as far as bather comfort and health is concerned, it is the air quality that is arguably of greater importance. However, this is intimately related to water quality, specifically the relationship between the disinfectant chemistry and the pollutant load.

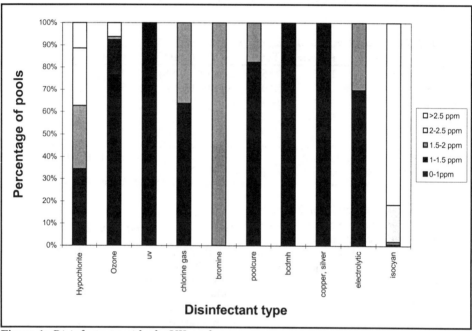

Figure 1: *Disinfectant residuals, UK pools*

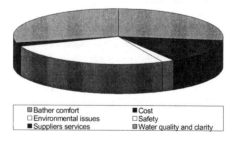

Figure 2: *Disinfectant choice*

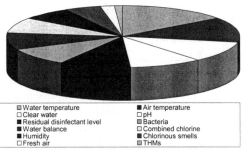

Figure 3: *Key pool quality aspects*

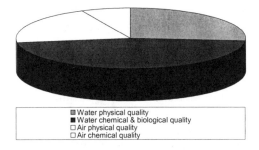

Figure 4: *Summary of Fig. 3 data*

2   THE CRANFIELD EXPERIMENTAL STUDY

Figure 5: *Pilot-scale plant at Cranfield*

Research into swimming pool water treatment has been conducted on both the bench and pilot scale at the School of Water Sciences at Cranfield University over the past six years. Much of this has been sponsored jointly by the Engineering and Physical Sciences Research Council (EPSRC) and the Pool Water Treatment Advisory Group (PWTAG). The most recent work has been based on a 2.2m^3-capacity pilot plant (Figure 5), a one-seventh linear scale model of an actual operating pool, incorporating all conventional unit

Table 4   *BFA composition*

Component	Concentration[†] g/l
NaCl	17.14
Na_2HPO_4	8.62
Na_2SO_4	6.14
$NaHCO_3$	2.24
KCl	9.26
$CaCl_2$	1.32
$MgCl_2$	1.77
NH_3	4.08
Urea	29.6
Creatinine	3.62
Histidine	2.42
Hippuric acid	3.42
Uric acid	0.98
Citric acid	1.24

1 bather = 25ml BFA

[†]50% dilution employed in later experiments

operations for swimming pool water treatment. The study has been based on the use of a body fluid analogue (BFA, Table 4), comprising the key dissolved organic and inorganic components of human perspiration and urine. include Trials have focused on the monitoring of both organic (THMs) and inorganic (chloramines) disinfection by-products (DBPs). Results for normal operation and shock loading have been obtained for a pool operated with conventional chlorination alone. Other techniques explored have included bromination, UV irradiation, reactive carbon and ozonation, the latter three being combined with downstream chlorination so as to leave a residual. Magnetic treatment was investigated on the bench scale only. All other treatment processes other than sand filtration were absent, thus effects recorded could reasonably be attributed to the disinfection mode rather than any other part of the system.

Key results obtained thus far for chlorine dosing comprise:

- for continuous dosing of BFA the change in DBP levels on changing key operational determinants by ± 50% was shown to be significant only on increasing the BFA dose rate (33% increase in THMs) and decreasing the disinfectant dose (50% decrease in THMs), the CA levels being largely unaffected throughout;
- intermittent dosing of BFA produces higher levels of THMs than continuous dosing at the same overall daily rate (Figure 6);
- the equilibrium ratio of chlorine arising in dissolved chloramines to that in dissolved THMs is around 42:1;
- a mass balance on the steady-state system can be carried out according to an assumption of either zero breakpointing or complete breakpointing to define the limits of DBP accumulation
- the zero-breakpoint mass balance accounts for 38% of the organic carbon, 61-63% of the chlorine and 84-~100% of the ammoniacal and amino nitrogen, the exact figures being very sensitive to assumptions made regarding trichloramine formation;
- the complete-breakpoint analysis accounts for 52% of the chlorine for 100% oxidation of the ammoniacal and amino nitrogen, the organic carbon usage being unchanged at 38%.

Results for other techniques indictate that:

- Overall DBP removal followed the order ozonation ~ reactive GAC > UV irradiation>chlorination>bromination, with the magnetic treatment results being inconclusive.
- Bromination generates about 5-8 times more dissolved THMs weight-for-weight than chlorination, with the ratio of the individual THMs changing from 4:1

CHCl3:brominated THMs for chlorination to 9:1 CHBr3:chlorinated THMs for bromination.

- The THM atmospheric concentrations for each halogenation method are then about the same.
- Reactive carbon behaves as a chemical reactor for chloramine destruction, with a first order rate constant of around 0.011 s^{-1} under the conditions tested in the study.
- Low-pressure UV irradiation at 30-120 mW s^{-1} cm^{-2} produces a decrease in the dichloramine level with a corresponding small absolute increase in the trichloromethane level.

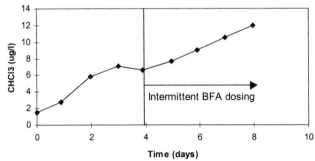

Figure 6: *CHCl3 levels arising from intermittent BFA dosing. Dosing with BFA up to four days was continuous and, under such conditions, the THM level stabilised*

3   CONCLUSIONS

Market trends generally indicate increasing use of chemical disinfectants, with %growth in pools at around 3.5% per annum throughout the early-to-mid 90's. A recent UK survey indicates that disinfectant choice is driven largely by bather comfort and water quality, these two being intimately related. However, research thus far at Cranfield would appear to suggest that the use of halogen-based chemicals results in an accumulation of chlorinated organic disinfection byproducts (DBPs) in the pool water, even though levels of the two most readily determined DBPs (trihalomethanes and chloramines) have apparently stabilised. The use of highly oxidative processes such as ozonation suppresses the formation of such products to some extent.

Pharmaceuticals, Cosmetics and Toiletries

A CURRENT LOOK AT PERSONAL CARE PRESERVATIVES

Michael Parkin

Business Manager, BU Biocides, Clariant Ltd

This presentation attempts to address the following questions:

- **What preservatives are being used?**
- **Why are they used, in what applications?**
- **What are the factors influencing choice?**
- **Does preservative use vary in different parts of the world?**
- **Preservative combinations**
- **Consumer awareness**

The scope of the discussion is described as follows:

Covers:
Personal care products
Leave-on & rinse-off cosmetics & toiletries
Colour cosmetics, hair care, baby & infant care
Skin creams / lotions, wet-wipes, sun-care
Bath & shower gels, hand & body wash, soaps

Does not cover:
Household products eg. fabric care, manual dishwashing, polishes
Antimicrobial 'actives' eg. antibacterial toilet soaps, facial wash, handsoaps

To address the first question, 'which preservatives are currently being used?', a very useful starting point is to examine the use of preservatives in the US market. This gives a fairly representative picture of the preservatives used, not only in the US market, but also in most other markets around the world with the notable exception of Japan. The graph below has been compiled from data from collated and published by D Steinberg, covering the most widely used preservatives in US cosmetic and toiletry products. It becomes very clear that the parabens esters, in particular the methyl and propyl esters, occur most frequently. Indeed it has been quoted that "after water, the most commonly occurring raw material in cosmetics is methyl parabens". As these 'top ten' preservatives are important to the personal care manufacturing industry, the key features, advantages and disadvantages have been listed in the following tables. To reflect the growing importance of a range of organic acids such as sorbic acid, dehydroacetic acid and benzoic acid, and which appear in the list below the 'top ten', a table summarising the relative strengths and weaknesses of organic acids is included.

FREQUENCY OF PRESERVATIVE USE

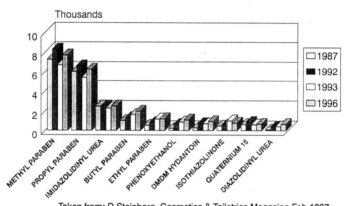

Taken from: D Steinberg Cosmetics & Toiletries Magazine Feb 1997

PRESERVATIVE	ADVANTAGES	DISADVANTAGES
PARABEN ESTERS **Typical use concentrations 0.1 - 0.3%**	Low toxicity Dermatologically safe at typical use concentrations Over 70 years widespread use with an extremely low incidence rate of skin response GRAS (Generally Recognised As Safe) in the USA Stable and active over a wide pH range Stable to heat Combinations of esters exhibit increased activity Approved for cosmetic applications worldwide	Low water solubility Some nonionics inactivate to varying degrees May require addition of other preservatives, eg. bactericides Incompatible with some proteins

PRESERVATIVE	ADVANTAGES	DISADVANTAGES
IMIDAZOLIDINYL UREA **Typical use concentrations 0.2 - 0.5%**	Water soluble	Poor antifungal activity
	Low oil solubility	Some formaldehyde release
	Good antibacterial activity	Relatively expensive
	Active between pH 4 - 9	Poor heat stability
	Non volatile	
	Easier to handle than formaldehyde	
	Low formaldehyde release (not activity dependant)	
	Activity potentiated with parabens	

PRESERVATIVE	ADVANTAGES	DISADVANTAGES
DIAZOLIDINYL UREA **Typical use concentrations 0.1 - 0.3%**	Water soluble	Weak anti-yeast activity
	Low oil solubility	Some formaldehyde release
	Good antibacterial activity	Relatively expensive
	Active between pH 4 - 9	Poor heat stability
	Non volatile	
	Easier to handle than formaldehyde	
	Low formaldehyde release (not activity dependant)	

PRESERVATIVE	ADVANTAGES	DISADVANTAGES
DMDM HYDANTOIN **Typical use concentrations 0.15 - 0.4%**	Cheap Water soluble Low oil solubility Broad spectrum of activity Active at low concentrations Active between pH 4 - 10 Non-volatile Good heat stability	Formaldehyde donor

PRESERVATIVE	ADVANTAGES	DISADVANTAGES
PHENOXYETHANOL **Typical use concentrations 0.4 - 1.0%**	Good activity against Pseudomonads Low toxicity Compatible with nonionics and proteins Broad spectrum combination with parabens	High concentrations required if used alone

PRESERVATIVE	ADVANTAGES	DISADVANTAGES
METHYLCHLORO-ISOTHIAZOLINONE **& METHYL ISOTHIAZOLINONE** **Typical use concentrations 7.5 - 15ppm**	**Broad spectrum of activity** **Active at very low concentrations** **Water soluble** **Compatible with nonionics**	**Skin sensitiser** **Maximum use level 15ppm** **Stability reduced above pH 8** **Used mainly in rinse-off products**

PRESERVATIVE	ADVANTAGES	DISADVANTAGES
ORGANIC ACIDS **BENZOIC ACID & Na SALT** **SORBIC ACID & K SALT** **DEHYDROACETIC ACID & Na SALT** **Typical use concentrations 0.2 - 0.5%**	**Very low toxicity** **Good fungicidal activity**	**Moderate bactericidal activity** **Low water solubility (acid)** **Activity mainly between pH 2 - 6** **Incompatible with cationics and some nonionics**

A number of other preservative systems are also widely applied to personal care products and are worthy of mention, some of these being used with increasing frequency, such as bronopol, methyldibromoglutaronitrile and iodopropynyl butylcarbamate, and some used with decreasing frequency, such as formalin and salicylic acid.

Given the range of chemicals which are available and permitted for use as preservatives, the question arises; why are some types very clearly favoured over others? To understand this, we have to look at the key factors which influence preservative choice. The most important of these are listed below:

- **Cost-effectiveness**
- **Product pH**

- **Global formulations**
- **Raw material compatibility**
- **Suitability for sensitive leave-on applications**

To a large producer of bulk surfactants for toiletries or a manufacturer of relatively low-cost bath and shower washes, the relative cost of the preservative is high. Hence to a typical user with this profile, the cost-effectiveness of the preservative may be one of the most important influencing factors. Under these circumstances, preservatives such as methylchloroisothiazolinone / methylisothiazolinone, bronopol and DMDMH tend to find acceptance.

The pH of the finished product may have a strong influence on the type of preservative used. A good example of this can be seen with the use of organic acids which may exist in a predominantly dissociated or an undissociated form as a consequence of the product pH. The undissociated form is considered to confer the antimicrobial activity and the effect of pH on benzoic, sorbic and dehydroacetic acid is described in the graph below. It can be seen that, at the normal pH of most personal care products ie. 5.5 to 7.0, there is little activity remaining. Hence organic acids would be suitable preservatives for predominantly acidic products, such as astringent washes made with lemons.

With increasing dominance of multinational personal care companies, there is an additional factor now to be considered when choosing the preservative; the need for the same product formulation to be sold on a global basis. Hence a large manufacturer with operations in many sites around the world may prefer to have one single formulation which can be produced at any, or all, of their worldwide factories. This means that the preservative must meet all local and national regulations. The chart below summarises the position of some of the most widely used preservatives in the three key markets of USA, Europe and Japan.

RELATIONSHIP OF UNDISSOCIATED ACID TO pH

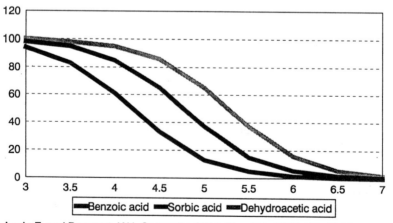

Luck, E. and Remmert, K-H. Cosmetics & Toiletries (Feb.1993) 108, 65-70

The raw materials used in a particular product may present some challenges to the formulation chemist when selecting the most appropriate preservative. Problem ingredients may act as microbial nutrients, preservative inactivators or preservative adsorbers and careful testing and assessment of the preservative system is necessary in these cases. One notable example of this is the inactivation of parabens esters by certain non-ionic surfactants.

APPROVAL IN USA, EU & JAPAN

PERMITTED USE CONCENTRATIONS

PRESERVATIVE	USA	EU	JAPAN
METHYL PARABEN	SAU	*1	*2
PROPYL PARABEN	SAU	*1	*2
IMIDAZOLIDINYL UREA	SAU	0.6%	750ppm***
BUTYL PARABEN	SAU	*1	*2
ETHYL PARABEN	SAU	*1	*2
CHLOROMETHYLISO - THIAZOLINONE	0.1%	0.1%	0.1%**
PHENOXYETHANOL	SAU	1.0%	1.0%
DMDM HYDANTOIN	SAU	0.6%	750ppm***
QUATERNIUM 15	SAU	0.2%	NO
DIAZOLIDINYL UREA	0.5%	0.5%	NO
SORBIC ACID	SAU	0.6%	0.5%
DEHYDROACETIC ACID	SAU	0.6%	0.5%

*1 Total paraben maximum of 0.8%   *2 Total paraben maximum of 1%
** Rinse-off products only   ***As free formaldehyde   SAU = safe as used

MAKING PRESERVATION EASIER

PROBLEM INGREDIENTS

Microbial nutrients
- Aloe
- Dipropylene glycol
- Plant extracts
- Sodium hyaluronate
- Vitamins

Preservative inactivators
- Cellulose gum
- Lecithin
- Non-ionics eg. Polysorbates
- Xanthan gum

Preservative adsorbers
- Kaolin
- Silica
- Talc
- Titanium dioxide

The skin compatibility of a preservative is a further consideration when, for example, formulating a hypo-allergenic skin care or baby care product. Whilst this is a subject of some debate and discussion with many published articles appearing in the scientific press, nonetheless there are some clear trends to be seen in usage of preservatives. This is most marked when comparing preservatives in 'leave on' cosmetics (such as conditioning creams, night creams etc) with 'rinse-off' products (such as shampoos, bath gels etc). It is a fair point that preservatives such as parabens, phenoxyethanol and organic acids tend to be used most prevalently in 'leave on' products, whilst preservatives such as DMDM Hydantoin, methylchloroisothiazolinone / methylisothiazolinone and imidazolidinyl urea are regularly formulated into 'rinse off' products.

A survey of product labels in the cosmetics section of a supermarket will quickly reveal that, quite rarely would a preservative appear in isolation. Preservative combinations are used very widely in most personal care products and there are some very tangible reasons for this:

- **To move closer to the ideal preservative**
- **To broaden the spectrum of activity**
- **To improve safety by reducing the concentrations required of any single component**
- **To be more cost effective / competitive**

There is a wide range of preservative combinations commercially available and some of the more important combinations are shown below

Preservative Blends

Typical blends currently available

- ## Parabens + phenoxyethanol
- ## Parabens + bronopol + phenoxyethanol
- ## Parabens + diazolidinylurea
- ## Parabens + benzyl alcohol
- ## Methyldibromo glutaronitrile + phenoxyethanol
- ## Methyldibromo glutaronitrile + bronopol + parabens + phenoxyethanol
- ## Methylchloroisothiazolinone + methylisothiazolinone + benzyl alcohol

There is a wide range of technical benefits to be gained by both the formulator and the consumer by using a preservative combination. A clear example of this is detailed in the following table. This demonstrates the well-known inactivation of parabens by non-ionics such as PPG-15 stearylether, ceteareth-20 and polysorbate-60. Combining parabens with phenoxyethanol may reduce the inactivation, whilst a liquid combination of parabens, phenoxyethanol and bronopol provided the best preservation. This example underlines some of the benefits of combinations, now widely recognised by most users. With increasing use of natural raw materials and ongoing pressure on reducing the concentrations of single preservative systems, the use of convenient, blended liquid preservative mixtures is likely to predominate in the future.

Preservative compatibility with non-ionic surfactants

Non-ionic	No inactivation	Significant inactivation	Complete inactivation
PEG-5 stearylether	Parabens Parabens + phenox. Parabens + bronopol + phenox.		
PPG-15 stearylether	Parabens + bronopol + phenox.	Parabens + phenoxyethanol	Parabens
Ceteareth-20	Parabens + bronopol + phenox.		Parabens Parabens + phenoxyethanol
Polysorbate 60	Parabens + bronopol + phenox.	Parabens + phenoxyethanol	Parabens

A PRACTICAL GUIDE TO THE SELECTION AND APPLICATION OF TOPICAL ANTIMICROBIALS, FROM A PRODUCT DEVELOPMENT PERSPECTIVE

Ali Altunkaya

Consultant, 4 Denne Road, Horsham, West Sussex RH12 1JE, UK

Antimicrobials are used at two levels: -

1) Preservatives
2) Actives

Preservatives-Introduction
Cosmetics and toiletries product manufacturers generally aim for 30 months shelf life and hence the product under development must have adequate preservative system to cope with the following: -

1) Selected **raw materials**, which may have organisms at the point of purchase. This is especially the case with natural raw materials produced with minimal processing to meet marketing claims;
2) **Compatibility** – Combination of raw materials is needed to obtain the desired effect from the end product by the user, at the same time minimising the possible reduction in activity of the preservative system;
3) The possible **contamination during manufacture,** filing and from the container used to deliver the product to the user;
4) To stand up to cold **temperatures** of the Canadian winter to Saudi Arabian climate during transit and storage, yet maintain its spectrum of activity
5) Likely contamination during **use;**
6) **Economical** and easy to use;
7) **Gentle** to the end user;
8) **Positive marketing claims** for the preservative if that is possible.

This is a tall order and this cannot be achieved for every product. However, the aim is to achieve as many of the above objectives with respect to preservation as possible.

The Preservative Selection
There are more than **70 preservatives** allowed in cosmetics in the EU and of those there are ten preservatives, which represent 90% of the volumes used (CTMS April/May 1993, by Dene Godfrey of Nipa Laboratories).

The preservatives I select for my formulations is much more limited than the ten listed in the above reference. The products which I develop for a company with natural ingredients image, restricted me to the **para hydroxy benzoic** acids, **benzyl alcohol** and **phenoxyethanol.** I have been allowed to use **Myavert C** during the last two years due to increase in customer complaints which was assumed to be due to the high levels of phenoxyethanol and benzyl alcohol used hence allergic reaction to the products. The Myavert C is made up of five components, and the original research was based on physiological defence mechanism found in most mammals, which help to protect mucous membrane. One component of the system, **Lactoperoxidase enzyme** is extracted from **milk whey** and this makes it a valuable tool to have a positive image with the marketers of cosmetic products.

Ingredients Selection Strategy for a Formulation: -
1) The **active constituents** selected for the formula to deliver the **positive benefits** comes first,
2) The type of ingredients, which will **hold and deliver** those ingredients onto and into the skin to give the desired effect to the user,
3) The **preservatives** which will give adequate safety to the user and the product, yet it will not be deactivated by the ingredients selected.

This should take into account the **solubility of preservatives** in different phases and **partition coefficient** (ref 2), the **effective pH** for the product and the preservatives (ref 3) in the combination. The manufacturers of these preservative compounds do not always provide the latter information. A range is quoted by the manufacturers where activity is observed, as though this is universal and that it has the same activity within all that range, irrespective of what you have in it.
Depending on which part of the body the product is aimed at would also influence the selection of preservatives as the mucous membrane, eye and face products require far more gentle ingredients than feet. Also, **leave on** products require more gentle preservatives than **wash off** products due to the increased contact time with the skin.

Irritation Potential and Preservative Performance
The use of personal care ingredients is increasing every year by about **2% in volume** and the potential allergic reactions to some of the ingredients are also increasing given the limited number of preservatives used within the industry. An article by Catherine Martin (Chemistry in Britain, March 1997 p35-38) quotes a Swedish study which shows that **12% of the population of Sweden** suffers from an **allergic reaction** to cosmetics: **32% of cases were attributed to preservatives,** 27% to perfumes, and 14% to emulsifiers. If that is translated to rest of the western world, which consumes the majority of the cosmetics and toiletries, potentially we have about 4% of the population who are allergic to one or more of the preservatives currently in use. This would generate massive customer complaints for the marketing companies, which is not observed in practice.

Selection of **preservatives,** in combination with **fragrance** and **emulsifiers** and other actives need to be evaluated together to minimise the potential allergy to the user. To that effect, the following combinations can help: -

1) Wherever possible I aim to incorporate **Myavert C, which** has one of the **lowest irritation potentials** in combination with **parabens** or **potassium sorbate.** This combination has reduced the customer complaints from a low to negligible level in all face products in which this combination is used by a skin-care marketing company. Typically the complaints for allergic reaction is in the range of one in 200,000 to 1,000,000 packs sold for face products suitable for all skin types. However, there are

limitations as to where this system can be used. One of these limitations is that it has to have an effective buffering system to keep the **pH within a 5.4-5.6** to prolong the shelf life and maintain its effectiveness at its maximum capability. (The manufacturers quoted effective pH range for this material is 5 to 7). **Gelling agent** selection will have to be made so that they are not affected by the ions used to create the buffering system. The final pH also restricts where it can be used as this reduces the number of **emulsifiers** available. The buffering systems also add to **material cost** and to **manufacturing cost**, as the specified final pH has to be tightly controlled. Where this is achieved the **positive benefits outweigh these disadvantages**.

The antimicrobial activity profile is as such it acts **very quickly** and in some products this is used to a positive effect as it enables an **antibacterial claim** to be achieved for face **cleansing and purifying products.**

2) Combinations of **Phenoxyethanol** and **Parabens-** This is used as a second option and where the pH can be controlled to ideally 6 and below and wherever possible restricting the amount of non-ionic surfactants. The total Phenonip type combination **not exceeding 0.7%**.

3) **Bronopol** (2-Bromo-2-nitropropane-1,3-diol) in combination with **methyl** and **propyl paraben.** The paper generated by the **Danish group** (ref. 5 Knoll BASF publication) shows that the mechanism of action of Bronopol is **independent of formaldehyde formation**. The inclusion of Tetrasodium EDTA at about 0.02% to remove ions in the water or extracts and carefully excluding any tertiary amines such as triethanolamine to counteract the fears of the marketers with reference to possible nitrosamine formation and release of formaldehyde. The pH used is usually at 6 and below.

4) Where high pH system is needed then a combination of **2,4-Dichlorobenzyl Alcohol** with **phenoxyethanol** is used such as Midtect PFP type mixtures from Microbial Systems International of Nottingham UK.

Natural Preservatives and Animal Testing
Preservatives which are based on natural ingredients or nature identical are small in numbers and these are gaining the attention of marketers. The ones I am aware of are: - **Grapefruit Seed extracts** and **Usnic Acid** distributed by Paroxite, **Neem seed oil** available from Microbial Systems International. The nature identical ones are **Myavert C** from BASF as mentioned above and Farnesol, which is supplied by Dragoco.

The Myavert C was chosen for inclusion in my development programme, because it had **no smell** and also it had the highest spectrum of activity against **the gram positive and the gram-negative bacteria** even compared to conventional anti-bacterials, yet it has **excellent safety profile.** This activity profile combined with **no animal testing** on this material makes this an ideal preservative for formulators and also for the marketers as they see the benefits from higher sales and lower complaints.

Anti-Microbials as Active Ingredients

Marketers are finding a **positive role** for anti-microbial compounds in areas which have not been seen before. Recent product introductions accompanied by major promotions by Savlon for 'First Aid-Hand Hygiene Liquid Gel' with claims such as 'kills 99.99% of germs in 15 seconds'. The main actives are ethanol, Triclosan and propylene glycol. Another major promotion from Dettol on Anti-septic properties is currently on our television screens, creating a positive image for use of these ingredients, which could only help this industry.

A paper given by David Ashworth on 'The use of Antimicrobials as Active ingredients from Head to Toe' during Preservetech 1996 summarises the traditional use of these actives for problem areas. Therefore I will only concern myself on areas, which are associated with marketers aiming to make cosmetic products with more positive antibacterial claims.

Cleansing and purifying products
Utilising the usual levels of anti-microbials for cleansing products with normal user instructions of three minutes or longer contact time can achieve positive claims of anti-bacterial for Myavert C based enzyme preservation system without the stinging associated with alcohol. Table 1 gives the plate kill speed data for a face masque where the preservative also becomes a positive attribute.

The same principle is used for face washes using Myavert C to cleanse the face of bacteria, without the stinging associated with other forms of anti-bacterial face wash.

Foot Cooling Products-
Using ethanol in combination with Bronopol at 0.05%, parabens and essential oils such as menthol an anti-bacterial claim was demonstrated as per table 2.

Analysis of table two shows that a staphylococcus aureus count of 1 million colony forming units per gram was killed off on plate within 5 to 15 minutes using very high levels of antimicrobials at a level only suitable for feet application whereas the Myavert C in the face mask achieved the same level of kill within three minutes, yet it is very mild and suitable for face and eye area application. Three minutes and longer application time for a product such as face is mask is common and this would achieve normal cleansing as well as microbiological purification of the face of the customer.

Conclusion
Other natural and nature identical preservatives with similar safety profile without the drawbacks of strong smell, strong colour, and short shelf life need to be developed to meet the consumer's needs. These would than gain the acceptance of the major multinational companies scientists and marketers, which would help to reduce overall allergy experienced from personal care products, and would enhance our lives and the standing of this industry.

Myavert C

The table 1 summarises the challenge test data using Myavert C 5 part enzyme based preservation in a face mask 1 including speed of kill test to justify positive marketing claims for the anti-microbial system used in the product. Table 2 uses conventional chemical anti-microbials for a foot product at much higher levels than would be allowed for face application yet the speed of kill of the enzyme system is superior yet much more mild.

Preservative Challenge Test Report   Table 1

Sample	Source		Sample Number		Date Tested	Test Number
Avocado and Pineapple Clay Masque			167/06		29.6.99	641/99
Batch 11 and 12 Ref 10 13.5.99						

Total viable count per gram   Bacteria   <10   Fungi   <10

Micro-organism	Baseline Count	Sampling Time	0hr	3min	5min	10min	6hr	24hr	48hr
Staphylococcus aureus NCTC 10788	1.80E+07	Count	5.50E+01	<10	<10	<10	<10	<10	<10
		Reduction factor	3.27E+05	NOR	NOR	NOR	NOR	NOR	NOR
Pseudomonas aeruginosa NCIMB 8626	2.10E+07	Count	4.00E+01	<10	<10	<10	<10	<10	<10
		Reduction factor	5.25E+05	NOR	NOR	NOR	NOR	NOR	NOR
Candida albicans NCPF 3179	1.40E+06	Count	1.45E+02	2.30E+02	8.00E+01	7.00E+01	<10	<10	<10
		Reduction factor	9.66E+03	6.09E+03	1.75E+04	2.00E+04	NOR	NOR	NOR
Aspergillus niger IMI 149007	1.70E+06	Count	5.40E+03	5.00E+03	4.40E+03	3.25E+03	<10	<10	<10
		Reduction factor	3.15E+02	3.40E+02	3.86E+02	5.23E+02	NOR	NOR	NOR

Micro-organism	Baseline Count	Sampling Time	7 day	14 day	28day	^Neutraliser Control	−Diluent Control
Staphylococcus aureus NCTC 10788	1.80E+07	Count	<10	<10	<10	180	184
		Reduction factor	NOR	NOR	NOR		
Pseudomonas aeruginosa NCIMB 8626	2.10E+07	Count	<10	<10	<10	54	68
		Reduction factor	NOR	NOR	NOR		
Candida albicans NCPF 3179	1.40E+06	Count	<10	<10	<10	85	31
		Reduction factor	NOR	NOR	NOR		
Aspergillus niger IMI 149007	1.70E+06	Count	<10	<10	<10	14	2
		Reduction factor	NOR	NOR	NOR		

Preservative Challenge Test - Kill Test   Table 2

Sample: Foot Cooler AA19 12.1.98   Test No.   340/98   Date: 1.4.98

Organism	Baseline	0min	1min	5min	15min	30min
				TVC		
S.aureus	9.60E+06	<100	<100	8.00E+01	<10	<10
C. xerosis	2.00E+05	6.00E+02	<100	1.00E+02	6.00E+01	<10
T. mentagrophytes	3.40E+04	7.00E+03	1.60E+03	2.00E+02	3.40E+02	3.60E+02

TVC on Receipt	
6.40E+02	Bacteria
<10	Fungi

References

1) Maintaining the Option for Preservation, by Dene Godfrey of Nipa Laboratory, CTMS April/May 1993 p 39-41
2) Significance of Partition Coefficient of a Preservative in Cosmetic Emulsion, by H S Bean et all, American Perfumer and Cosmetics, Vol. 85, March, 1970
3) The influence of pH, Emulsifier, and Accelerated Ageing upon Preservative Requirements of O/W Emulsions, by Gene Jacobs, M.S. at all Soc. Cosmet. Chem., 26, 105-117 (February, 1975)
4) The Art of Preservation by Catherine Martin, Chemistry in Britain – March 1997
5) Bronopol and Formaldehyde, October, 1995 publication of summary of papers by knoll BASF Ref. No. P1-2.09
6) The use of Antimicrobials as Active Ingredients from Head to Toe- by David Ashworth of BASF. Preservetech 1996 proceedings.

Subject Index